Fruit and Vegetable Carving

Meera John Jacob

photographs by
Harold Teo

TIMES BOOKS
International

First published in 1974 as
The Art of Vegetable Carving
by Federal Publications Sdn. Bhd.

This revised and enlarged edition
published in 1982 by
Times Books International

Designed by Jenny Soh

Cover Photograph: ONION DAHLIAS

Printed by: Tien Mah Litho Printing Co., (Pte.) Ltd.

038208

To
My Husband

PREFACE

I have long been fascinated by the art of fruit and vegetable carving which I find not merely pleasing and entertaining but also quite educative and utilitarian.

Many things can be carved out of vegetables, statuettes and other solid objects being but a few examples. Nothing, however, can equal flowers made out of vegetables for beauty, originality or versatility, because in producing one flower not only can many vegetables be combined exquisitely, but the same vegetables may be coloured differently to produce a variety of altogether new kinds of flowers. A green rose, a blue marigold or a red jasmine—flowers simply non-existent in nature are within reach! For petals carved out from vegetables may be dyed any colour or combination of colours according to one's whim and fancy. As a creative art, flower creations out of vegetables has few equals.

This art is also not without its utilitarian value. To the houseproud entertainer, it gives a unique method of beautifying the dining table. Children are greatly excited by the ample opportunity to discover for themselves new kinds of flowers and other articles of a decorative nature. For the professional cook, fruit and vegetable carving, with its unlimited scope for novelty and display, will help secure laurels and greater glory and reputation for his institution.

The greatest value of the art of carving will be realised when flowers are scarce or out of season. On such occasions, vegetables are a veritable godsend. They are also excellent substitutes, better ones at that, for while the natural flower has to be thrown away after use, those made from vegetables may be eaten after they have been served as a feast for the eyes.

In the following pages will be found plenty of helpful hints and advice for those interested in this art or anxious to cultivate an interest in it. To the beginner especially, this book is a *must;* the professional, I am sure, will find it stimulating. I therefore recommend this unhesitatingly and with great pleasure to one and all.

CONTENTS

1 ONION

12 Lotus Bud
13 Lotus Flower
14 Yellow Lotus
15 Chrysanthemum
17 Water-lily
18 Sunflower
19 Dahlia
20 Onion Rose
21 Cosmos
23 Onion Lily

2 RED RADISH

26 Eight Flowers

3 TURNIP/YAM BEAN

32 Turnip Rose
33 Candlestand
34 Anthurium

4 EGGPLANT

38 Daisy
38 Earth Star

39 Gerbera
40 Chinese Bell Flower

5 TOMATO

44 Tomato Tulips
44 Tomato Rose
46 African Daisies
46 Tomato Beetle

6 WHITE RADISH

50 Cherry Blossoms
51 Radish Rose
52 Magnolia Flower
54 Calla Lily
54 Clematis
55 Two Peonies
58 Narcissus
60 Radish Daisy

7 CARROT

64 Spider Chrysanthemum
65 Daffodil
66 Poppy
68 Ginger Flower
68 Carrot Rose

8 CUCUMBER

70 Two Patterns
71 Cucumber Fingers

9 CHILLI

72 Red Bells
73 Japanese Witch-hazel
73 Chilli Daisy

10 CAPSICUM

74 Capsicum Daisy
75 Capsicum Dish
75 Fire Lily

11 PAPAYA

78 Papaya Basket
79 Papaya Vase
79 Papaya Boat

12 ORANGE

80 Orange Basket
81 Orange Lotus

13 PINEAPPLE

84 Pineapple Chicken
85 Pineapple Container

14 WATERMELON

88 Melon Basket

15 GRAPEFRUIT

89 Grapefruit Dish

16 PRACTICAL HINTS

90 Basic Equipment
94 Care of Equipment
94 Choice and Storage
95 Preservation of Flowers

1 ONION

Onions are easy to carve; however, it is advisable to smear your hands with a few drops of vinegar before cutting the onions. It will save your eyes from irritation and your hands from smell. Make sure that the onions you select are single-bulbed, and not divided in two or more sections as some onions are.

Flowers carved from onions will last for about a week if you keep them in polythene bags in the refrigerator when not in use. When required for use again, soak the flowers in ice-cold water for about 10 minutes. Flowers carved from onions are ideal for flower arrangement and are a good substitute when natural flowers are not available or not in season.

Note: Before carving any onion, peel off the dry outer skin. Trim, *but do not cut off,* the root portion which normally forms the base of the carved flower. Wash the onion before use.

Lotus buds in an arrangement with the Sacred Lotus and lotus leaves in a low container which acts as a miniature lotus pond.

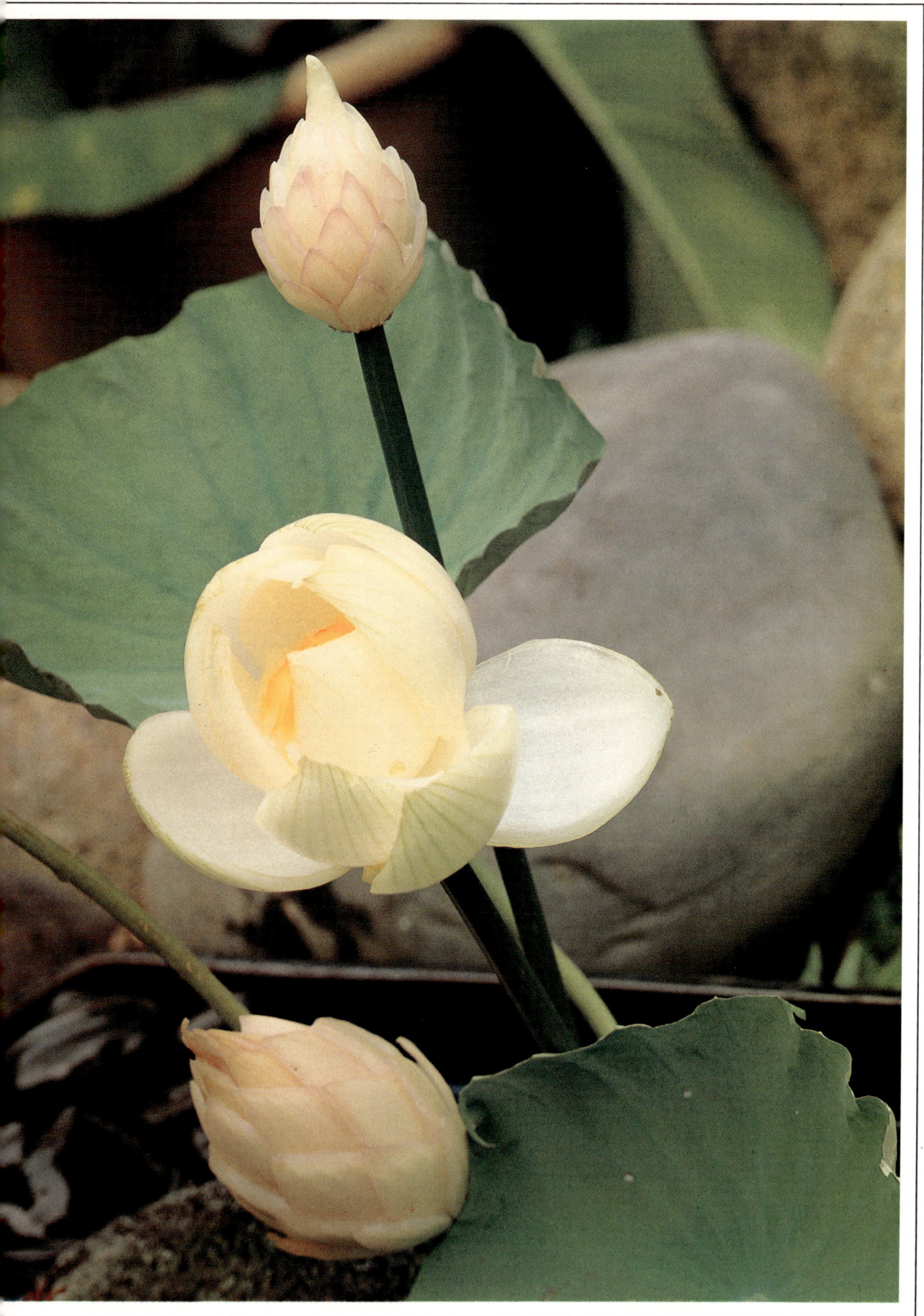

LOTUS BUD

Illustrated on page 11

Select an onion (shaped like a bud rather than round), white or purple. Remove the dry outer skin. Cut horizontally 2½ cm from the bottom of the onion. Do not allow the knife to cut into the second layer. After cutting, remove the top part of the first layer. This operation will be clear from a look at **Figures A** and **B**. About 1 cm above the first layer, cut through the second layer and take away the top part. If the size of the onion permits, cut into the third and fourth layers also in a similar way. Finally, cut triangular wedges on each layer working from the base upwards, as shown in **Figures C** and **D**. Lotus buds are used with lotus flowers in an arrangement.

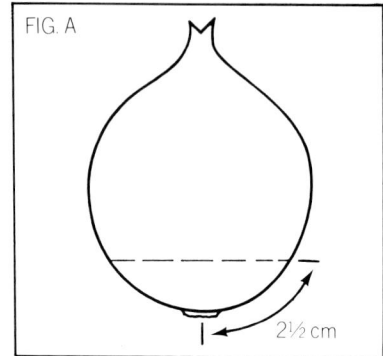

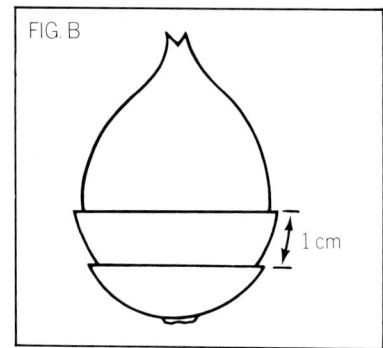

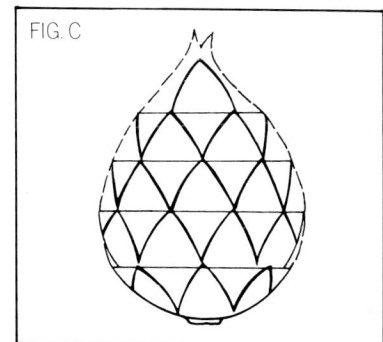

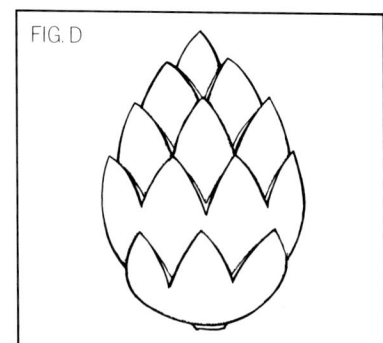

FIG. A

2½ cm

FIG. B

1 cm

FIG. C

FIG. D

LOTUS FLOWER (Sacred Lotus of India)

Illustrated on page 11

An elegant flower to carve is the lotus. Use a large bulb-shaped onion, white or purple. Remove the dry outer skin. Cut three petals from the outer layer, as shown in **Figure A**. Do not allow the knife to cut into the second layer. Gently peel open the petals. Between these three petals, cut three more petals from the second layer as in **Figure B**. Gently open up all the petals to form the flower as shown in **Figure C**. Soak the flower for 30 minutes in ice-cold water.

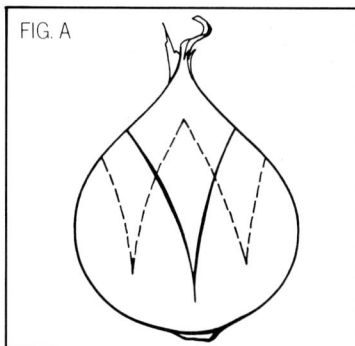

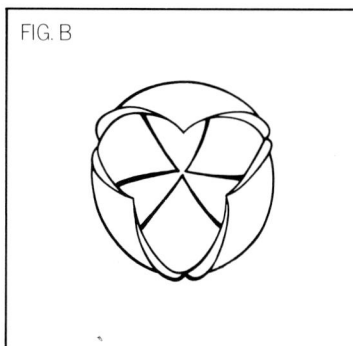

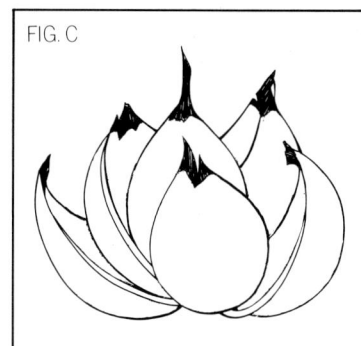

FIG. A

FIG. B

FIG. C

LOTUS POND

Illustrated on page 11

This is a decorative arrangement of lotus buds and flowers in a miniature lotus pond which will add glamour to the dining-room.

For the leaves, trim off any large leaf to a circular shape to resemble the lotus leaf. If you have lotus leaves in your garden select a few small leaves together with the stems. Use cypress grass stems for the flower stalks, and use cocktail sticks to fasten buds and flowers to stalks.

Choose a large, flat container and pour some water into it. Cover the pin holders with some pebbles or crab shells.

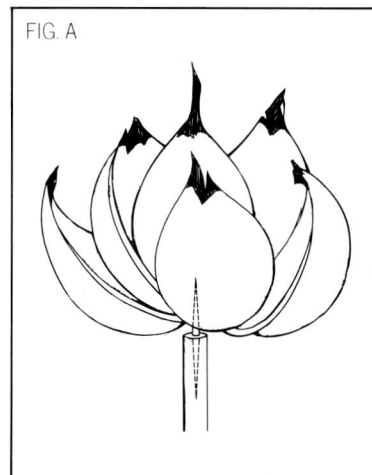

FIG. A

The Yellow Lotus, exquisite even sitting simply on a large leaf.

YELLOW LOTUS

Illustrated above

To carve the yellow lotus which has many petals use a large, plump, white onion. Make eight vertical cuts from the top downwards, leaving about 1 cm at the bottom as in **Figures A** and **B.** Gently open up each petal. Soak the flower in ice-cold water for 30 minutes. The flower may be left white or coloured yellow with a vegetable dye.

You can also make a lotus with six vertical cuts, but it will give fewer petals.

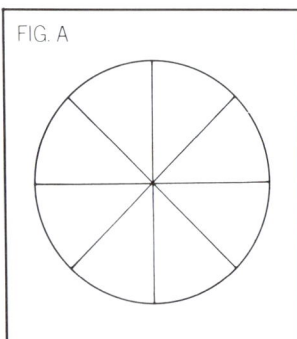

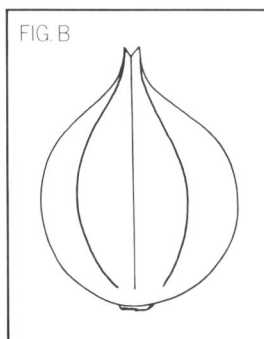

Small leaves with Chrysanthemums: contrast in size and shape for an attractive decoration.

CHRYSANTHEMUM

This flower can be made with a white or purple onion. Remove the dry outer skin. Make as many vertical cuts as you can, making sure that the vertical cuts are all perpendicular and do not intersect one another except at the centre of the onion as shown in **Figures A** and **B**. Carefully peel the petals open and insert a cocktail stick through the base for attaching to a stalk as in **Figure C**. Soak the flower in ice-cold water for 30 minutes.

This onion chrysanthemum will serve as a splendid component for flower arrangement. If made with a white onion, it may be dyed to any desired colour. This flower may also be used to decorate a cooked dish.

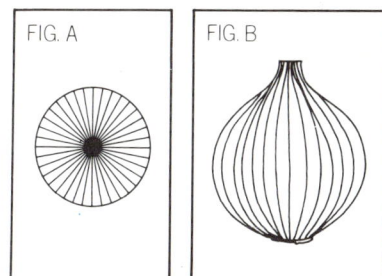

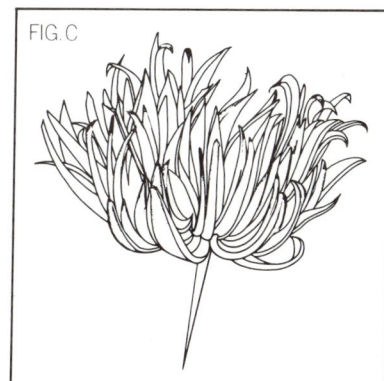

FIG. A

FIG. B

FIG. C

15

Water-lilies set in a miniature pond lend a cool refreshing atmosphere and help lighten a gloomy corner.

Select a white or purple onion as desired. Using a knife with a thin blade, make triangular cuts to resemble the tips of water-lily petals. The cuts should go right through to the middle, as shown in **Figures B** and C. Gently separate the two parts. Both parts can be used as water-lilies.

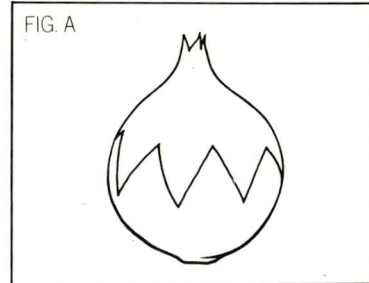
FIG. A

The second way of setting this flower is done by cutting the onion into two parts horizontally. Take the top portion and separate each layer except the central layers, very carefully. Use three or four of the outer layers and cut triangular wedges around the edge of each layer. Arrange the layers in order of size (largest outside) and such that the tips of the petals of each layer come in between the petals of the next layer (**Figure D**).

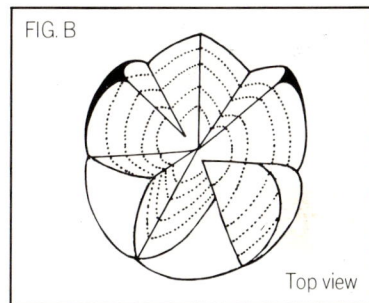
FIG. B
Top view

A central piece is required for this flower. For this, cut a round piece of radish about 12 mm high and make many vertical criss-cross cuts, leaving about 2 mm at the bottom of the piece uncut. Dye the central piece yellow and attach it to the centre of the flower with a stick, as in **Figure D**. Alternatively, use a piece of carrot for this purpose. This gives the whole flower a more natural look.

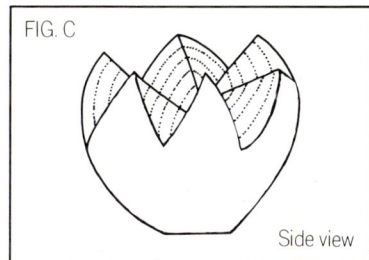
FIG. C
Side view

For a third variety, select a large, single-bulbed white or purple onion. Cut three or four petals from the outer layer, as in **Figure A** on page 13. Cut four or five more layers of petals, alternating the petals. Carefully cut off the remaining layers in the centre, leaving a short stump about 1 cm from the bottom. Make vertical criss-cross cuts on the stump and dye the stump yellow.

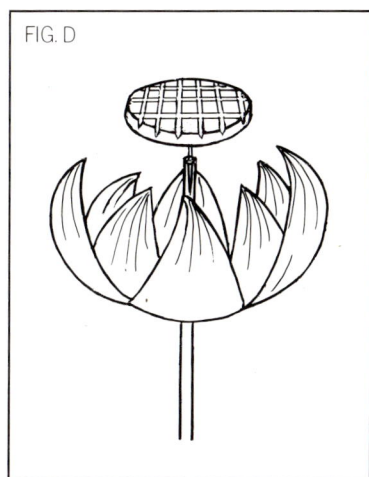
FIG. D

Note: If you wish to use the water-lily in a miniature pond, it should be placed on the surface of the water, since it is there that the water-lily blooms. Remember to soak the flower in ice-cold water for 30 minutes.

A burst of Sunflowers with New Zealand flax.

SUNFLOWER

Illustrated above

This is a most satisfying flower to carve, and just a few sunflowers will make a pretty, natural arrangement.

Peel off the dry outer skin of a big white onion. Cut the onion 2 cm from the top. Discard the top portion. Follow the directions on page 13 for carving the petals of the lotus, but the width of each petal should be only 1 cm. Carve a second layer so that the petals of the second layer show between the petals of the first layer.

Using a small knife, cut off the centre of the onion, leaving a large round stump 2 cm from the bottom. Make several vertical cuts on the stump and slowly open the strips a little. Dip the whole flower in yellow vegetable dye. Either dot the stump with brown dye, or gently press some mustard seeds into the stump.

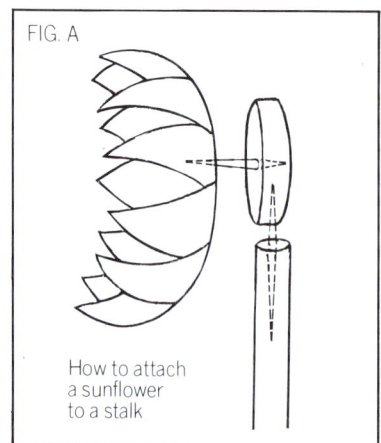

FIG. A

How to attach
a sunflower
to a stalk

Showy Dahlias take on an air of elegance arranged with slender twigs and buds kneaded from stale bread.

DAHLIA

Illustrated above

The petals of a dahlia are not as long as those of the sunflower but the carving procedure is similar.

Select a large, single-bulbed white or purple onion and peel off the dry outer skin. Cut off 3 cm from the top of the onion. Carve out petals as directed on page 18 for the sunflower. For the dahlia, however, do not stop at the second layer, but continue carving each layer until the centre is reached. Soak the flower for 30 minutes in ice-cold water.

If a white onion is used, dye the flower your favourite colour.

19

Roses you can eat.

ONION ROSE

Illustrated above

Select a bulb-shaped white onion. Slice off about 3 cm from the top of the onion. Check, by looking at the cut, that it is single-bulbed. If it is not, start with another onion.

Start carving the first layer of the onion, taking care that the tip of your knife does not cut into the second onion layer. Carve petals from the first layer to resemble rose petals (see **Figure A**). Remove the onion portions between the petals. Carve petals from the second layer, bearing in mind that the petals of one layer should appear between the petals of the next, outer layer. As you reach the centre and it is not possible to carve any more petals, make a few vertical cuts and dye the centre yellow. The onion rose makes a pretty garnish for cooked food.

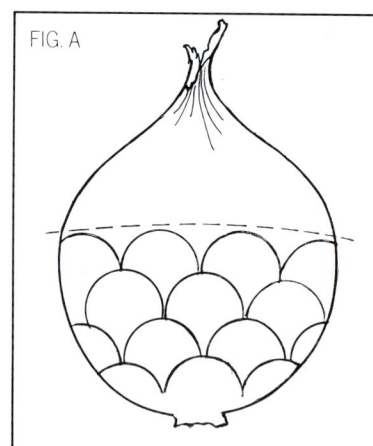

FIG. A

An original Christmas table with Onion Cosmos on the tree, Carrot and Turnip Candlestands, even a Turnip Rose.

COSMOS

Illustrated above

For this flower (and the next), choose a long, narrow single-bulbed onion as the carved result should be small, unlike the big sunflower.

Cut eight petals in the first layer of the onion (see **Figure A**). The width of each petal should be only 1 cm and the length 3 cm. With a small knife, cut off the rest of the onion inside, leaving a ½ cm stump as the centre of the flower. Leave the petals white and paint the centre yellow with a brush.

The cosmos can be used for flower arrangement by attaching the flower to a cypress stem with a cocktail stick. A few flowers with some leaves in a basket also make an attractive centrepiece for a table.

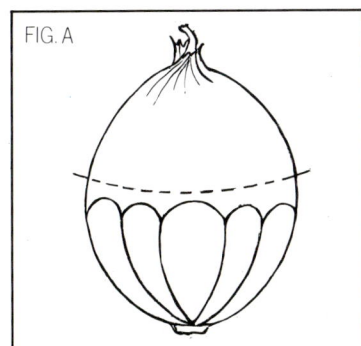

FIG. A

FIG. B

21

Several Onion Lilies carved from just one onion.

The onion left over from making a cosmos can be used to make several onion lilies. With a sharp knife, cut six petals of 1 cm width around the onion. The petals should be long and tapering (see **Figure A**), and cut only to about 2 cm from the tip.

Hold the tip of the first layer in one hand and slowly twist the portion inside two or three times to remove it (see **Figure B**). Carve the next layer of petals from the portion removed. Many flowers of gradually decreasing size can be made in this fashion till the centre is reached. In the photograph opposite, all the lilies were carved from one onion.

Dye the lilies any colour, but first peel off the thin waxy membrane covering each layer so that the colour spreads evenly.

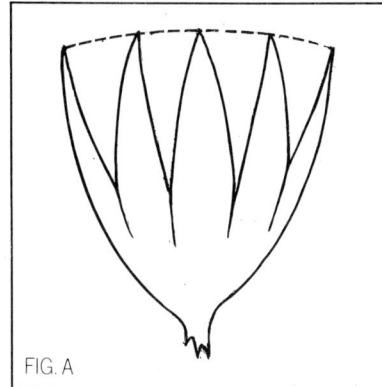

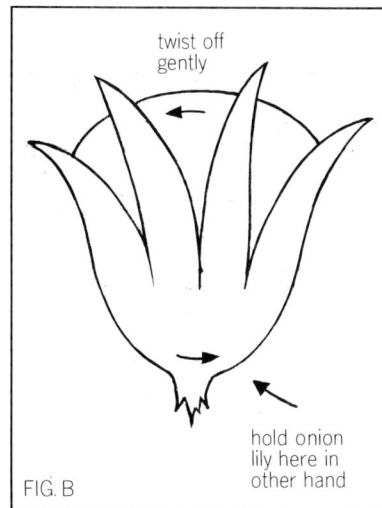

FIG. A

twist off gently

hold onion lily here in other hand

FIG. B

23

2 RED RADISH

This vegetable is the edible root of the species *Raphanus Sativus* and is usually eaten raw.

Since it is solidly homogeneous, it can be cut easily into any shape or design required, while its beautiful colour makes it ideal for decorative purposes. It is especially useful to adorn simple dishes such as salads and sandwiches. Parsley or coriander leaves go well with radish flowers. In this section nine different flowers are shown carved from the radish. Except for Radishes 1, 2 and 7, the flowers should be kept in ice-cold water before use. This serves two purposes: it keeps the flowers fresh as well as causes the petals to spread naturally.

A paring knife is best for carving radish.

A pretty country basket with an assortment of Radish Flowers.

RADISH 1

Illustrated on page 25

Select a round radish and cut away equally spaced wedges from the top to about halfway down, as shown in **Figures A** and **B** and use the flower for decorating cooked dishes.

FIG. A

Side view

Top view

FIG. B

RADISH 2

Illustrated on page 25

Choose a fairly large bulb-shaped radish and slice away a small piece near the root of the vegetable as in **Figure A** right. Scoop out a little flesh from the inside. Wedge the petals to about ¼ cm thickness and carefully remove the alternate skin. The result can be seen in **Figure B.** At the bottom of the radish (on both sides), make two V-cuts and remove the wedges.

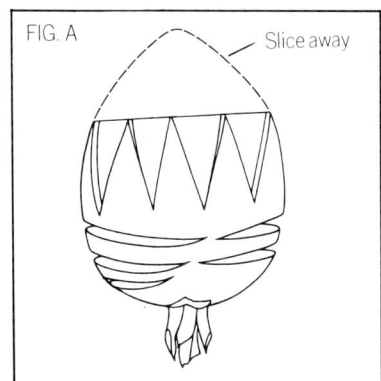

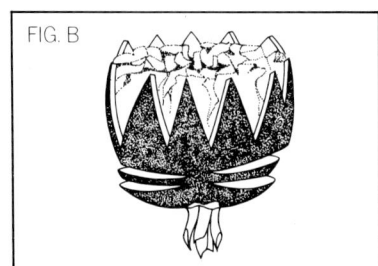

FIG. A

Slice away

FIG. B

26

RADISH 3

Illustrated on page 25

Cut vertical slices about ¼ cm thick from top to bottom, and slice on both sides. Cut off root. Place in ice-cold water before using as a decoration for cooked food. (See **Figures A** and **B,** right.)

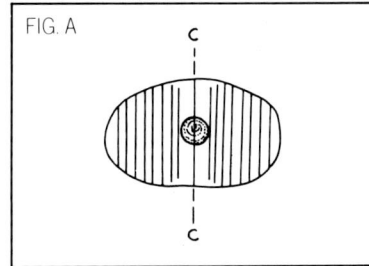

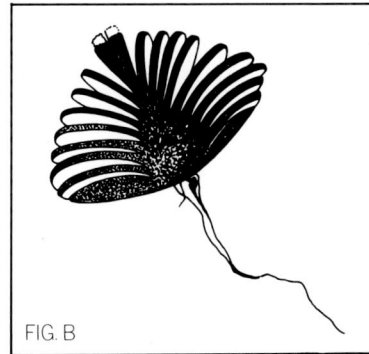

FIG. A

FIG. B

RADISH 4

Illustrated on page 25

Chionodoxa

Choose a bulb-shaped root. Cut five petals of about ¼ cm thickness to form the outer layer, as in **Figure A**. Clean the flesh by removing the leftover red skin. Between the petals of the outer layer, cut five more petals from the flesh to the same thickness. The remaining cone-like piece may be carved to a smaller size or removed carefully. Place the flower in ice-cold water before using to allow petals to spread out. (See **Figures A** and **B** at right.)

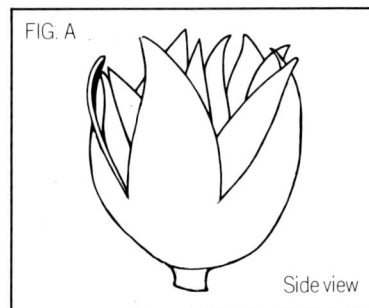

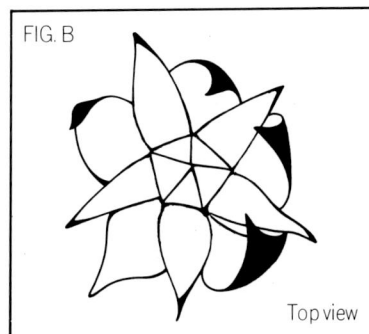

FIG. A

Side view

FIG. B

Top view

Chrysanthemum

Select a big radish and make many vertical cuts about ¼ cm apart from the top to about three-quarter way down. Turn the vegetable around and make another set of vertical cuts perpendicular to the first set, to form ¼ cm squares as in **Figure A** right. Place the flower in ice-cold water for petals to spread out (see **Figure B**). This flower looks very attractive when used to decorate dishes like fried rice, noodles or cakes.

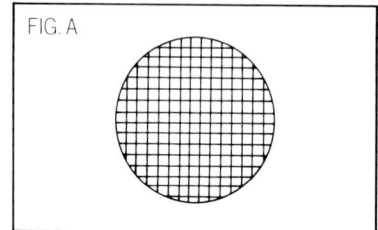

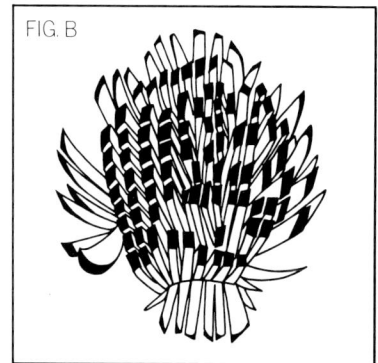

FIG. A

FIG. B

Take a bulb-shaped radish. Make five or six petals as shown in **Figures A** and **B**. Clean the inside and carefully carve the centre portion so that it resembles a round, white ball. Place the flower in ice-cold water and the petals will open out a little. As with the other radish flowers, use this flower for decorating cake or fried food.

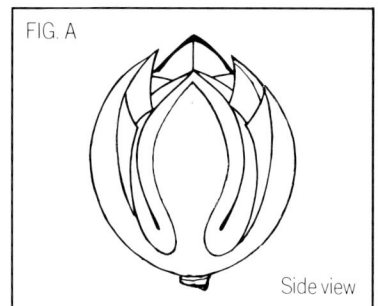

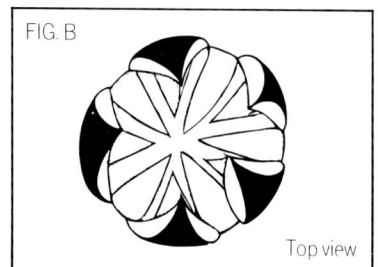

FIG. A

Side view

FIG. B

Top view

RADISH 7

Illustrated on page 25

Radish Cup
Select a large radish and slice away the top part as in **Figure A** right. Scoop out the flesh with a scoop, leaving a cup-shaped vegetable about ¼ cm thick. Cut a zig-zag pattern round the edge as shown in **Figure B**.

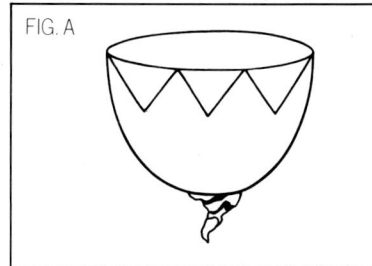

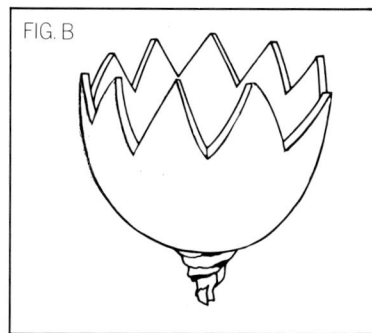

FIG. A

FIG. B

RADISH 8

Illustrated on page 25

You will need a big radish for this flower. Cut two slices, each about ¼ cm thick, on four sides to about 1 cm from the base, as shown in **Figure A**. Halve these petals vertically to get sixteen smaller petals. On the top of the central piece cut four wedges as shown in **Figure B**. Cut off the root. Place the flower in ice-cold water and use it as a decoration for cooked dishes.

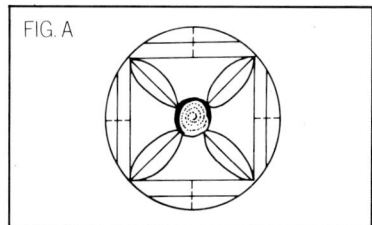

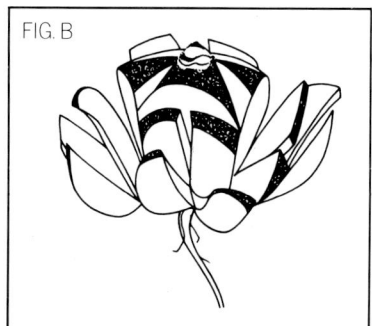

FIG. A

FIG. B

3 TURNIP/ YAM BEAN

Both the turnip and yam bean are edible tuberous roots and spherical in shape. The yam bean is known locally as *sengkuang or bangkuang*. The skin is brownish and the flesh white, crunchy and slightly sweet. It is often eaten raw or in a salad. The turnip and yam bean are almost identical in appearance and flesh texture and both may be used for the flowers in this chapter. They are easy to carve and ideal for flowers with large or stiff petals.

A thoughtful touch to a serving of crisps with a Turnip Rose.

Choose a long (rather than the usual round) turnip for this flower. Cut out a portion about 6-7 cm long. Peel and wash it. Mark a face with seven or eight sides on the broader end as in **Figure A**. Trim down the length following this figure to obtain a seven- or eight-sided block. Carve the block to taper it at one end to a cone. Slice thin petals down each side, as in **Figure B,** up to about 1 cm from the thin end of the cone. The petals will overlap naturally. Shape the petals with a small knife or a pair of scissors. Carve out a little of the flesh before beginning the next layer so that the layers of petals are not too close. Now carve a second layer of petals to come in between the first, outer layer. Proceed in this fashion till the centre is reached. Dye this turnip rose any favourite colour.

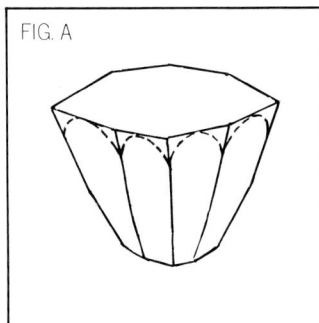

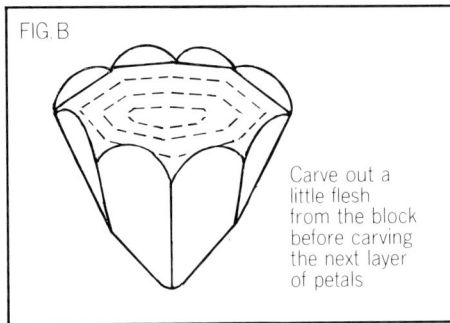

FIG. A

FIG. B

Carve out a little flesh from the block before carving the next layer of petals

Two rustic Candlestands you'll be proud to own.

CANDLESTAND

Illustrated above

Select a long turnip. If a big stand is required, choose a big turnip. Slice all around to obtain a smooth long piece with one rounded end and one flat one so that it can stand.

Leaving a circle of 2 cm diameter at the top of the rounded end, cut small triangles with ½ cm sides around the circumference, making sure they are evenly spaced. The bases of the triangles form the circumference of the circle. With the tip of a small knife wedge off the flesh within the triangles. Now cut diamond shapes between the triangles (see **Figure A**). Remove flesh within the diamonds. Proceed till the base is reached.

Leave the candlestand white or dye it the colour of your candle. A large carrot may be carved in the same way to give a pretty orange candlestand as in the photograph above.

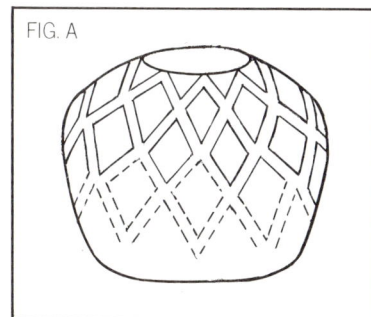
FIG. A

33

ANTHURIUM

Approximately a quarter of the turnip or yam bean is required to make one flower. So, from one large turnip or yam bean, at least four flowers can be made. Scrape off the outer skin and cut the root into four equal sections as in **Figure A**. Scoop out the flesh from the inside leaving about 3 mm around as in **Figure B**. This becomes the petal of the anthurium. From a 'scrap' piece, scraped off earlier, shape a small piece, as in **Figure C**, which will act as the pistil. Scratch a few vertical lines on the inner side of the petal to give it a natural look (see **Figure D**).

Use a cocktail stick to attach the pistil to the base of the petal as shown in **Figure E**. The other end of the stick is inserted into the stem of a cypress grass. Dye the petal red or yellow and the pistil yellow with vegetable dye. It is not necessary to place the flower in water.

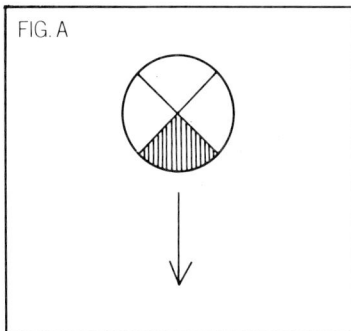

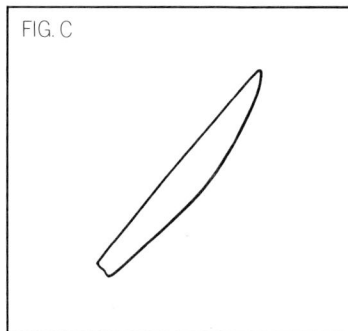

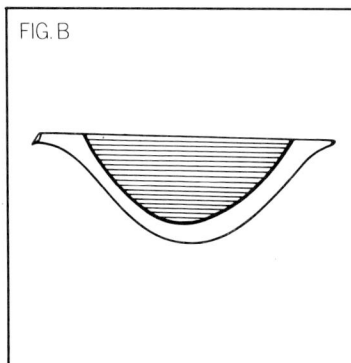

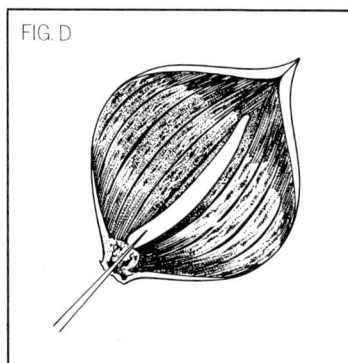

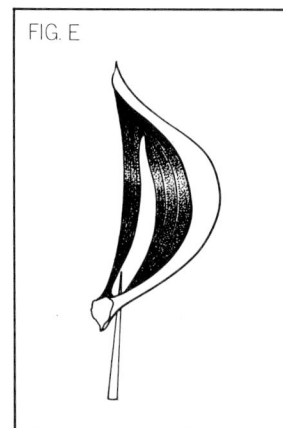

FIG. A

FIG. C

FIG. E

FIG. B

FIG. D

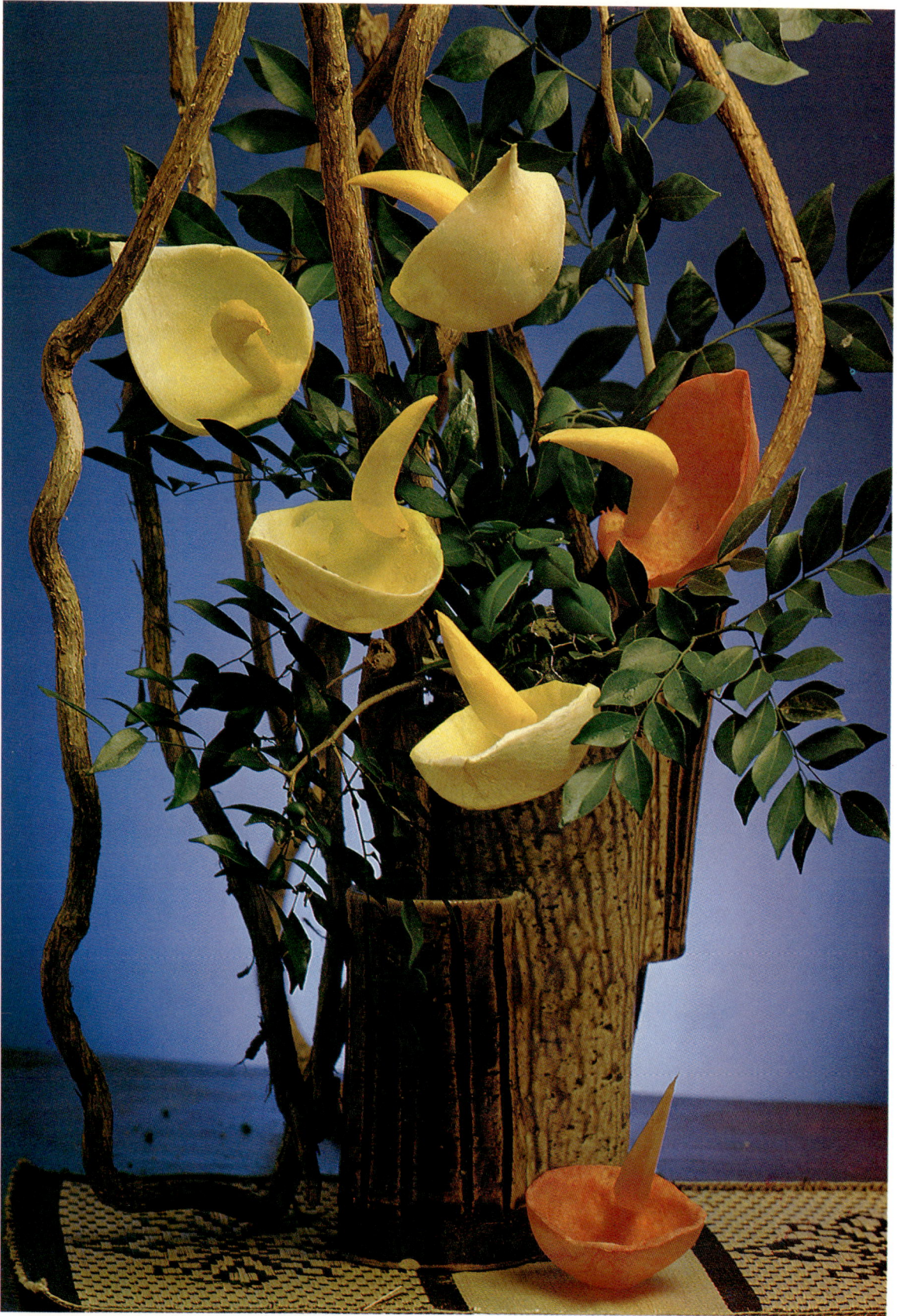

Anthuriums, red and yellow, in an arrangement.

4 EGGPLANT

Of the several varieties of eggplants (brinjals) two are commonly grown in local vegetable gardens. One is oblong in shape and purple. The other is usually egg-shaped and may be white or light green. Though they have little nutritive value, eggplants are nevertheless a popular vegetable used in curries and sambals and are reasonably cheap to buy and readily available in the market.

Daisies, Earth Stars and Gerberas . . . there's more to eggplants than mere vegetable for curry!

DAISY

Illustrated on page 37

Select a long purple eggplant. Cut it across about 7-8 cm from the stem portion, as in **Figure A**. Scoop out flesh, reserving a ball of flesh. With a pair of scissors, cut many vertical strips about 3 mm apart (see **Figure B**). If desired, attach the reserved ball scooped out earlier to the centre. Soak the flower in ice-cold water for an hour. When the petals of the daisy open, dye it any colour and attach it to a cypress stem with a cocktail stick. A coconut palm (*satay*) stick may be used instead of a cocktail stick and cypress stem. Sheath the coconut stick with a spring onion leaf.

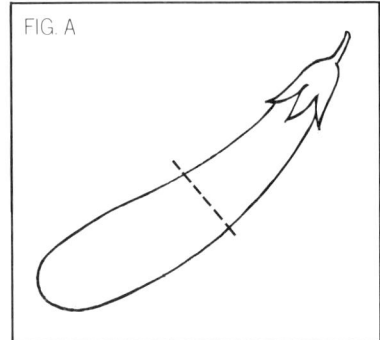

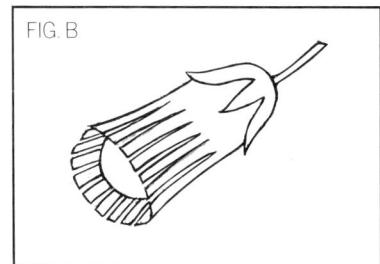

FIG. A

FIG. B

EARTH STAR

Illustrated on page 37

Take the other end of the eggplant used for making a daisy. Cut off a portion to obtain a piece 9 cm long (from the rounded end to the lateral cut). Mark six divisions on the face of the cut as in **Figure A**. Cut vertically through the mark to about 2½ cm from the bottom. Mark out the shape of the petals (see **Figure A** again) and cut petals, discarding flesh in between. Cut off flesh under the petals, leaving only the skin as the petals of the earth star.

Cut a slice off the rounded end to enable the earth star to stand. Attach a ball (scooped out earlier from the eggplant) to the centre of the earth star with a cocktail stick. Dye the earth star any colour.

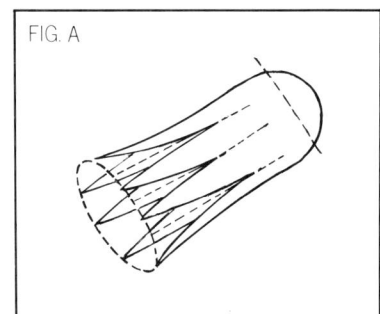

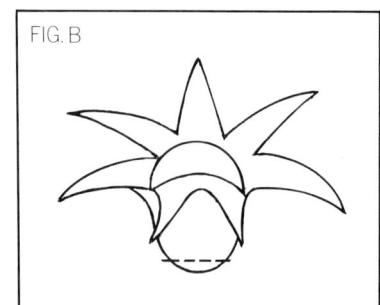

FIG. A

FIG. B

GERBERA

Illustrated on page 37

Choose a long round eggplant. Cut away the calyx and slice off the rounded top, leaving a 5 cm long portion. With a small sharp knife, carve petals of 1 cm width all round (see **Figure A**). Slice the inside all round to smoothen the solid portion within and carve a second layer of petals so that the petals of this layer appear between the petals of the outer layer. Carve the solid portion left to get a smooth bulb for the centre. Dip the whole flower in yellow, red or purple dye and refrigerate till all the petals curl up, giving the flower a natural look. The gerbera is an ideal flower for decoration.

Exercise: Try to make a gerbera out of a cucumber from the instructions above. A cucumber daisy and bell flower can also be made from a fat cucumber.

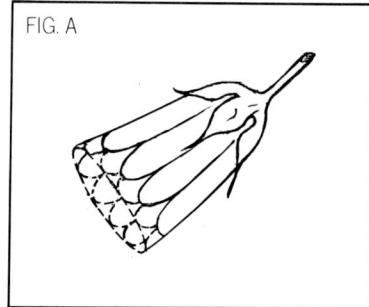

FIG. A

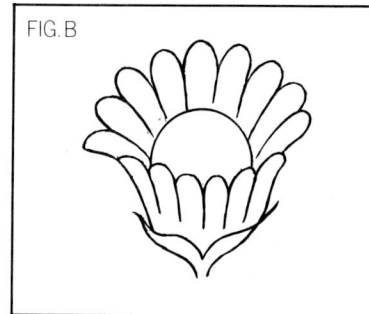

FIG. B

CHINESE BELL FLOWER

Illustrated opposite

Select a round or oblong eggplant according to the size of the flower desired. If an oblong eggplant is used, halve it laterally and use the portion with the stem. Cut five petals on the skin as shown in **Figures A** and **B.** From the flesh, cut five more petals about 3 mm thick which will come in between the outer petals, as in **Figure C.** Carve the remaining part of the inside into the shape of a ball. Cut off the stem at the base to make it stand firmly on a tray or plate if you want it to decorate cooked food. But keep the stem if you are going to use it for flower arrangement. Eggplant flowers for flower arrangement should be attached to cypress grass stems with cocktail sticks.

Note: After making the flowers, soak them in salt water for 10 minutes to prevent discoloration of the vegetable. Soak the flowers in dye solution and refrigerate immediately till ready to use, but not longer than 24 hours.

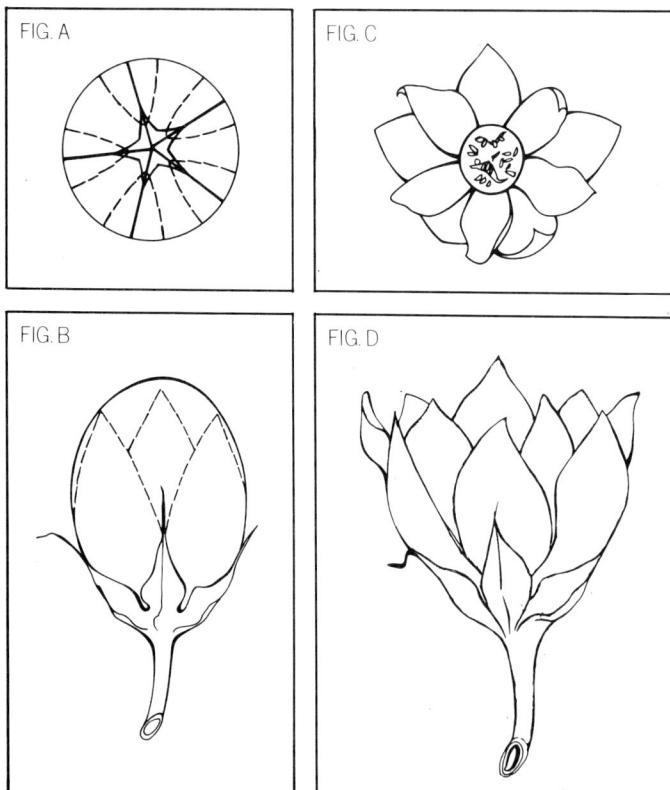

FIG. A

FIG. C

FIG. B

FIG. D

Large Chinese Bell Flowers are well balanced with large leaves.

5 TOMATO

Always use fresh and firm tomatoes for vegetable carving. The flowers should only be made about one or two hours before they are needed. All the tomato flowers suggested here are ideal either for flower arrangement or for decorating cooked food.

Tomato Tulips and tomatoes—both equally luscious.

TOMATO TULIPS

Illustrated on page 43

Select a few fresh, firm tomatoes which are oval in shape. Invert the tomato so that the base is on your palm. Make two cuts crossing each other as shown in **Figure A**. The cuts should be about halfway down the tomato. Use a small scoop to clean the inside, leaving about 3 mm of the flesh beneath the skin. With a pair of scissors, carefully cut the four portions, shaping them into petals as in **Figure B**.

Use cocktail sticks to attach the flower to any stem, or use coconut palm (*satay*) sticks and sheath them with spring onion leaves.

Make a few tulips in this way and use thin long leaves and twigs or dried materials for an arrangement in a vase or basket.

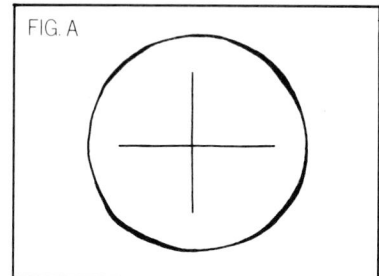

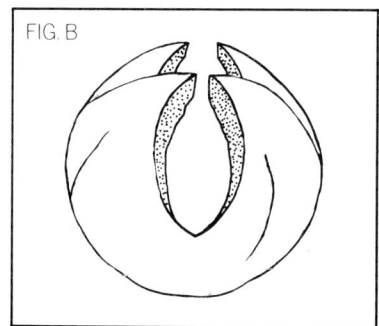

FIG. A

FIG. B

TOMATO ROSE

Illustrated opposite

Slice the tomato at the base, where the calyx and stem are, with a very sharp knife, but do not cut off the entire portion. This slice becomes the base of the flower. Continue slicing the skin together with a thin layer of the flesh underneath, starting from the base and proceeding around the tomato. The strip should be about 12 mm wide. The flower is formed by winding the peeled skin round the base. This tomato rose is used to decorate dishes like fried rice, the Chinese 'cold dish', or any fried food. Parsley, coriander leaves or real rose leaves will go well with the tomato rose.

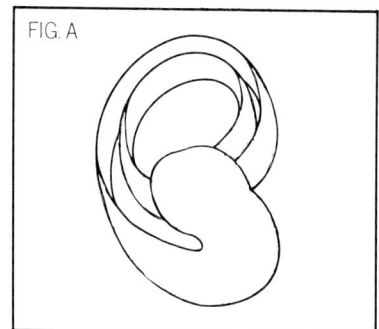

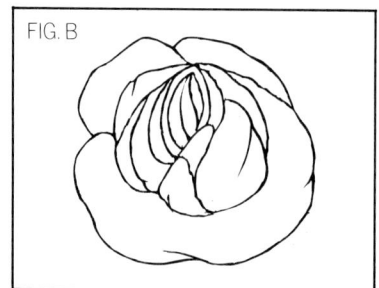

FIG. A

FIG. B

For a place setting with a personal touch, a Tomato Rose takes next to no time to complete.

AFRICAN DAISIES

Choose a fairly large and firm tomato. Holding it in an inverted position as in **Figure A** right, cut many vertical strips, each about 3 mm wide. Open up each strip to form a petal. Leave the inside of the tomato which resembles the pistils and stamens as in **Figure B**, right. African daisies may be used for flower arrangement.

A different version of an African daisy is made by slicing off about 1 cm from the top of a round tomato before proceeding as instructed above. This version can be seen in the photograph opposite.

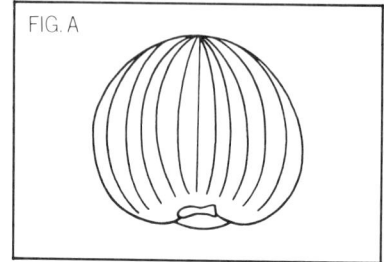

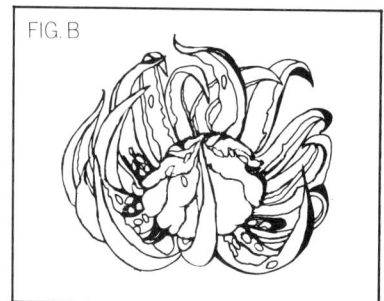

FIG. A

FIG. B

TOMATO BEETLE

Cut out a quarter segment of a tomato for each beetle. The skin portion will constitute the back of the beetle. Slice off about 7 mm of the core as in **Figure A** to secure a proper 'seat' for the insect. To obtain wings, gently slice under the skin up to about 1 cm from one end, the 'head' of the beetle. Trim skin to the shape shown in **Figure B** and cut down the middle using a pair of scissors so wings can be moved as shown in **Figure C**. Use a toothpick to make holes where the eyes and feelers should be. The eyes and feelers are made with thin strips of cucumber skin. Carefully push the eyes and feelers into place as in **Figure C**.

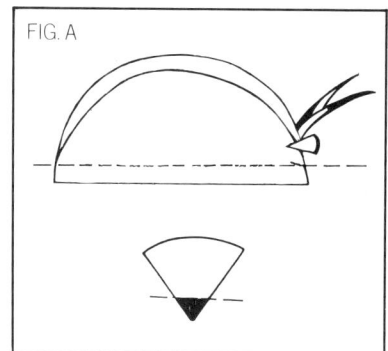

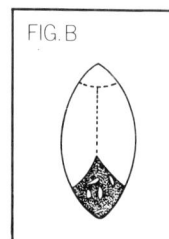

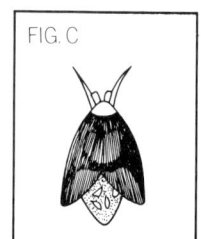

FIG. A

FIG. B

FIG. C

Thrill the children at your next party with fruit cocktail in a Papaya Boat and a basket of African Daisies.

6 WHITE RADISH

Another variety of radish which is white, this vegetable is normally long in shape and easily available in the market. When choosing a white radish ensure that the vegetable is not soft or crushed due to rough storage. They should be firm and not sound hollow when struck with the back of the finger.

The very accessories for an oriental arrangement of Cherry Blossoms—two Japanese fans!

CHERRY BLOSSOMS

Illustrated on page 49

Select a fresh radish but avoid using one which has been kept refrigerated. Using a sharp knife, cut out a length of about 5-7 cm from the broader end of the radish if you wish to make large flowers and from the narrow end for smaller flowers. The portion cut out should preferably be straight. Use toothpicks to mark six points as shown in **Figure A**. Carve away some flesh in between the toothpicks. The cross-section will now appear as in **Figure B**. Remove the toothpicks. Now cut curved slices about 3 mm thick (see **Figure C**). This can be done by first cutting off a little flesh from either end of the length of radish so that the face of each end is curved slightly.

For the centre of the flowers, cut small round pieces from a beetroot, purple eggplant or red radish, and attach a piece to the centre of each flower with starch, paste or a small piece of fine wire. An easy alternative is to paint in the centre using red vegetable dye.

To arrange the flowers attractively, cut any branch from a plant—a gardenia plant, for example—and remove all the leaves from the branch except for a few here and there. Using a pin holder, fix the branch and keep it in position in a vase. Using pieces of fine wire of about 5 cm length, tie each flower to the branch.

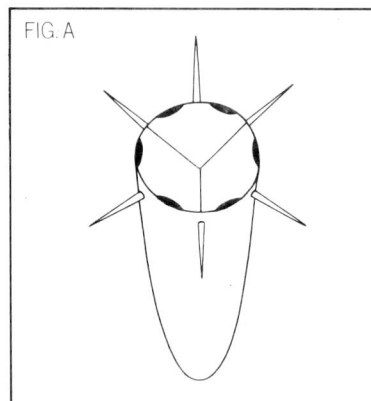

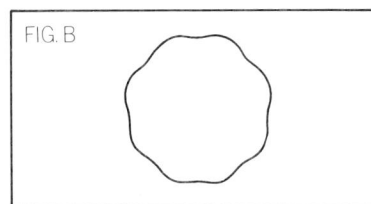

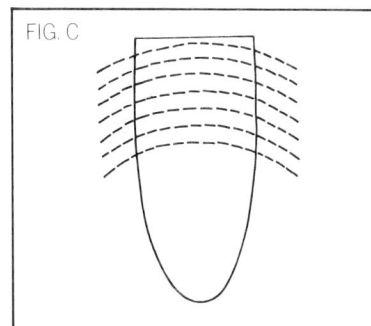

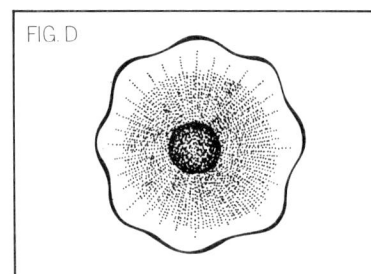

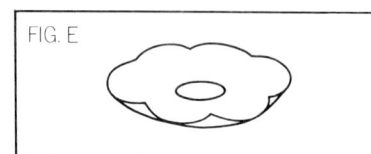

FIG. A

FIG. B

FIG. C

FIG. D

FIG. E

50

RADISH ROSE

Illustrated on page 83

Select a large and fresh long radish. Scrape off the skin and wash. Using the slicer, slice out a few round slices of about 3 mm thickness. Put the slices in a plate and sprinkle some salt on them to make them soft and pliable. Take one slice and roll it. This will be the innermost petal. Add each additional petal on to the slice, one at a time, as in **Figure A.** If you want a big flower, you can use more than eight slices. The slices in the form of a rose can be held together with a rubber band around the base. (See **Figure B.**) Wash the flower thoroughly in water and dye it any colour using vegetable dye. This is done by pouring a few drops of dye in a cup of water and immersing the flower in the solution for about 30 minutes. After this time, transfer the flower onto a saucer and refrigerate it until needed. Slice off a little portion carefully from the base so that it will stand firmly on a plate when you decorate cooked food with this flower. Use real rose leaves to add to the decoration.

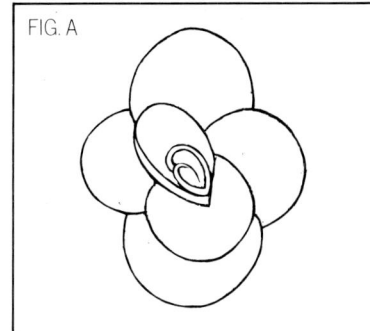

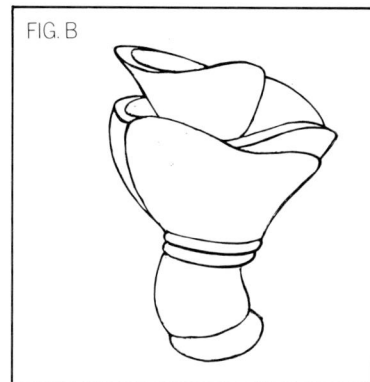

FIG. A

FIG. B

Select a large, white radish and cut out a section of about 12½ cm in length from the broader end of the radish. Scrape off the skin. Slice the portion lengthwise into slices of about 3 mm thickness. Select five big slices to be made into the petals of the flower.

Using scissors or a sharp knife, cut each slice to the shape of a petal as in **Figure A.** At the base of each petal make a straight cut about 2 cm long as in **Figure A.** Next, take a purple onion with a diameter equal to about half the length of the petal. Make a chrysan-themum out of this onion (see page 15). Keep it in ice-cold water so that the petals will open out. This will be the centre of the magnolia flower.

Cut out a stopper of any shape from the leftover radish. Insert a cocktail stick through the stopper. Attach the petals one by one as shown in **Figure B** to form the flower. Place the onion chrysanthemum in position at the centre as in **Figure C** and the flower is complete.

Magnolia flowers can be attached to cypress stems for flower arrangement. These flowers look attractive in an arrangement for the dining-table.

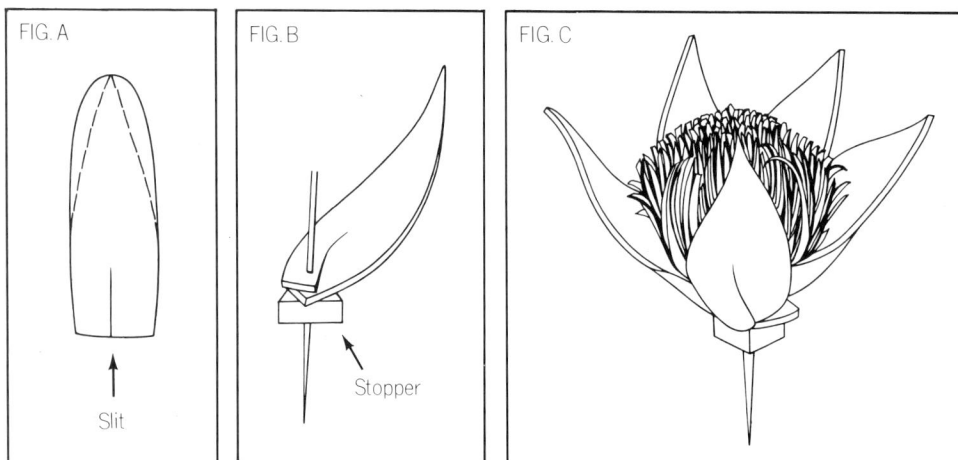

FIG. A

Slit

FIG. B

Stopper

FIG. C

A solitary Magnolia bloom for a centrepiece.

CALLA LILY

Illustrated on page 61

Select a large radish and cut it diagonally into two equal parts. From each part cut out a few slices of about 3 mm thickness as in **Figure A**. These slices are for the petals of the flowers. At the base of each petal make a slit about 1 cm long, as in **Figure B**. From the remainder of the radish carve a pistil for each petal and dye it yellow with vegetable dye. Wrap the base of each petal around the pistil and hold it in place with a stick or pin as in **Figure C**.

As an alternative, take a cocktail stick and insert half of it into the cypress stem. Attach the pistil to the other half of the cocktail stick. Hold the tapering end of each petal and fold it around the base of the pistil. The petal is held to the pistil by the use of a pin.

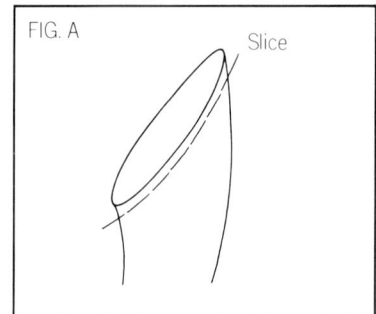

FIG. A   Slice

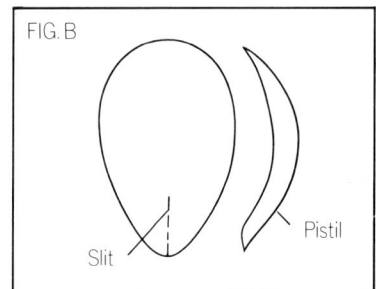

FIG. B   Pistil   Slit

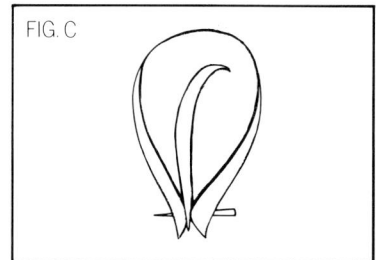

FIG. C

CLEMATIS

Illustrated on page 87

Scrape and wash a radish. Cut off a portion about 10 cm long. Carve the sides to obtain a five-sided block tapering at one end like a cone. (See **Figure A**.) Slice down the five sides, *not the edges*, as in **Figure A**, up to about 1 cm from the narrow end. Carefully, by curving the knife inwards, cut off the central stump and shape the petals.

Dip the flower in a diluted blue dye, then immediately take out the flower and dip it in diluted red dye to obtain the correct clematis colour. Clematis also occurs in white.

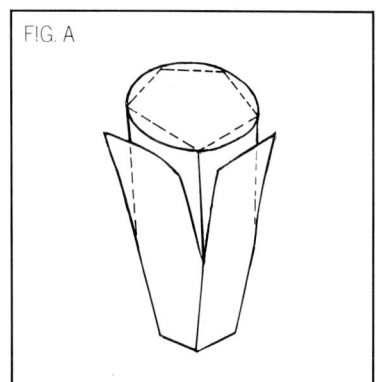

FIG. A

A fine water spray on Peonies and ferns gives them a fresh, just-picked appearance.

PEONY 1

Illustrated above

Select a medium-sized carrot and cut out a 7 cm length from the middle. Carve down the length to obtain a cone, tapered to a narrow end (see **Figure A**). Make wedges on the broad end, as in **Figure A** at right. Using a small knife, make five curved slits, downwards and evenly spaced, as in **Figure B**, about ½ cm from the wedged broad end. Cut another two circles of five slits each, alternating the slits.

Select a white radish of roughly the same size as the carrot. Scrape and wash it. Slice laterally to obtain five small, five medium-sized and five large rounds.

Insert the radish rounds into the carrot slits, using a small knife. The small rounds go into the topmost circle of slits, i.e., nearest the wedged broad end, the largest ·rounds into the last row of slits. Immerse the whole flower in ice-cold water for 30 minutes before using for an arrangement.

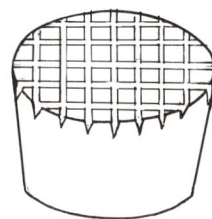

FIG. A

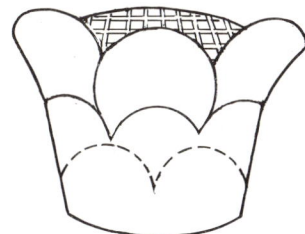

FIG. B

Select a long white radish which is slightly curved. Cut out a section about 10 cm in length from the curved part so that the cross-sections at the two ends have approximately the same diameter. Scrape away the skin and wash the radish.

From both top and bottom ends slice away portions of about 2 cm and 12 mm respectively as in **Figure A**. Hold the radish with the ends curving downwards, leave about 2 cm width along the centre of the top surface and slice off two strips lengthwise of about 1½ cm on both sides of the top surface.

Make wedges on the two faces as shown in **Figure B**. The pattern of the wedges may vary and some examples are shown in **Figure C**. **Figure D** shows how the petals are carved out of a single section of the radish. The petals carved out would be of varying sizes similar to the one shown in **Figure E**. The central portion of each petal may be tapered so that it is thinner than the ends.

Cut out a round section of carrot, about 2 cm thick, and wedge the top surface as shown in **Figure F**. This is to be used for the centre of the flower. Scoop out half a ball of carrot and use this as a stopper.

Insert a cocktail stick right through the stopper and arrange the petals on top of this, beginning with the largest petal and ending up with the smallest one as shown in **Figure H**. Then insert the central piece of the flower, the wedged carrot. The other end of the cocktail stick can be attached to a cypress stem if the flower is used for an arrangement.

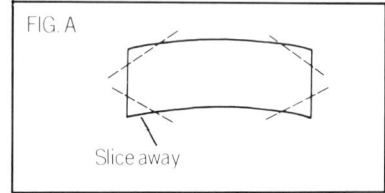

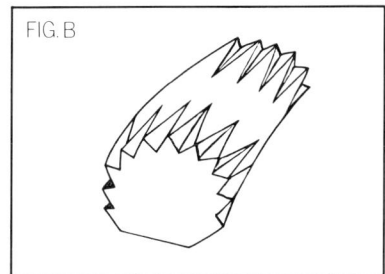

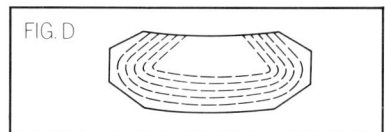

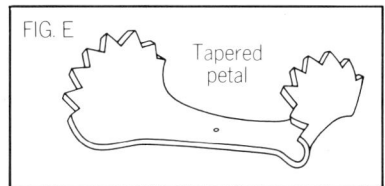

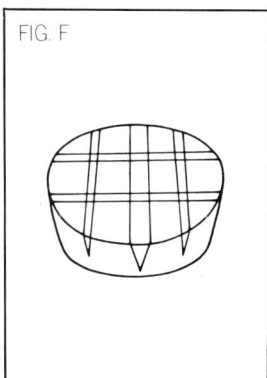

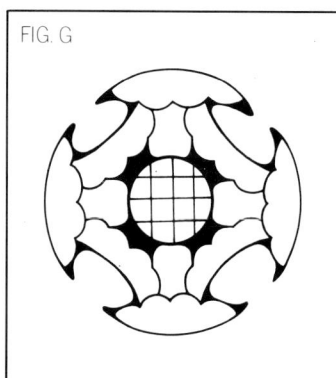

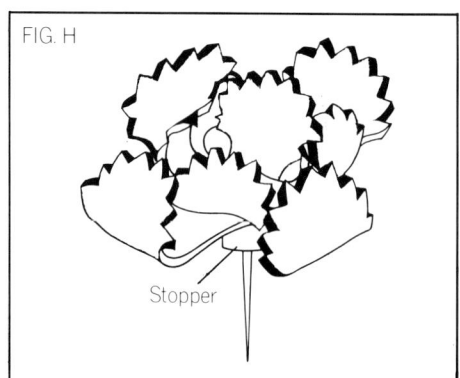

FIG. A

Slice away

FIG. B

FIG. C

FIG. D

FIG. E

Tapered petal

FIG. F

FIG. G

FIG. H

Stopper

A bamboo stand, a few sprigs of leaves and one large peony on a dish—simple yet effective.

Select a radish or carrot. After washing it, scrape off the skin. Cut a section about 12½ cm long from the broader end and slice six pieces lengthwise. Each slice should be about 3 mm thick. Stack them together and cut petals the shape given in **Figure A**.

Take a carrot and cut out a small round section about 7 mm thick from it. Wedge on one side as in **Figure B**. This will be the central piece of the flower. Now scoop out a ball from the same carrot. This is to be used as a stopper behind the petals. Push a cocktail stick right through the ball of carrot and arrange the petals one by one starting with the outermost petal as shown in **Figure C**. Cut off the stick protruding from the stopper. Place the wedged central piece in position. Take another cocktail stick and insert it vertically into the stopper as in **Figure C**. The other end of the stick is inserted into a cypress grass stem.

Arrange a few of these flowers in a low vase with tall grass or long leaves to decorate the living room.

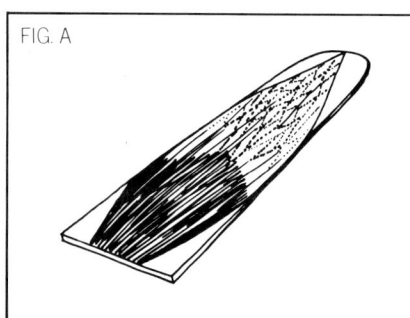

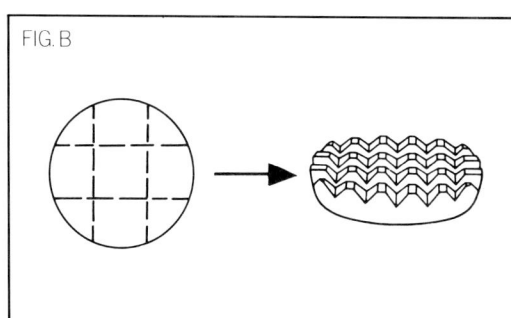

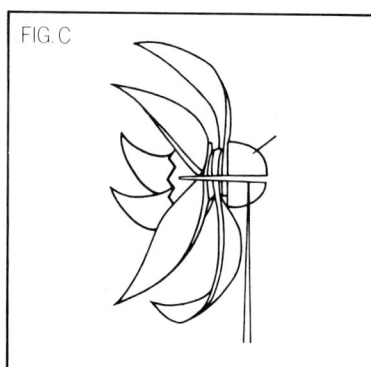

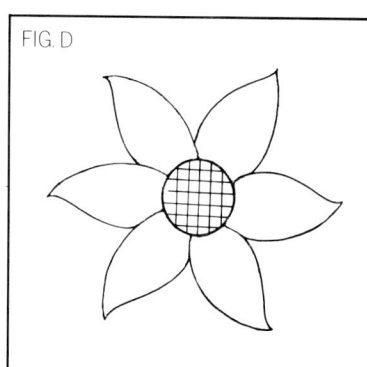

FIG. A

FIG. B

FIG. C

FIG. D

Not the easiest arrangement to attempt if space is limited, this requires a continuous line, highlighted by Narcissus and Daffodils.

RADISH DAISY

Illustrated opposite

Cut a 6 cm length from a large white radish. Check that the inside is perfectly solid and smooth. Slice all round to get an even cylindrical surface. With a sharp knife, slice all round as far as you can without breaking the piece. The object is to obtain a 6 cm wide, ¼ cm thick piece of radish, of any length. (See **Figure A**.)

Spread this long piece on a cutting board and cut 5 cm long strips on one side, leaving an uncut 1 cm space along the other side. Cut a 3 cm long portion from a carrot for the centre of the flower and shape to a cone (see **Figure B**). Make many cuts on the top surface as in **Figure B**. Wrap the long radish piece around the carrot piece and fix in position with cocktail sticks. Cut off the protruding ends of cocktail sticks.

Soak the whole flower in ice-cold water for an hour and keep refrigerated till the petals curl. Use the flower to garnish cooked food.

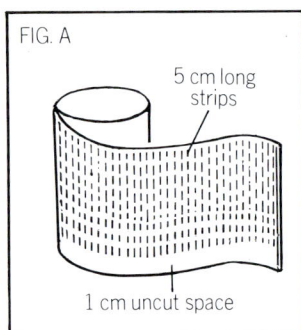

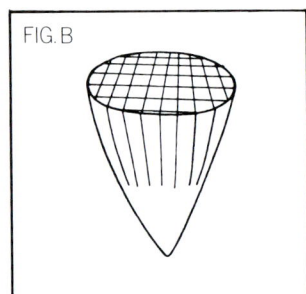

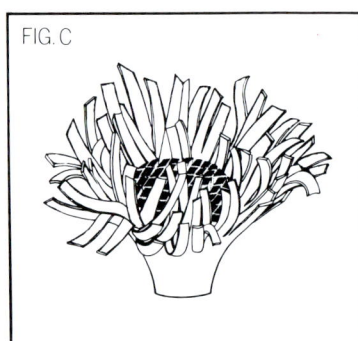

FIG. A

5 cm long strips

1 cm uncut space

FIG. C

FIG. B

Radish Daisies, Calla Lilies and Chilli Flowers make excellent food garnishes.

7 CARROT

Carrots are ideal for vegetable carving. Their natural rich orange colour makes flowers carved from them very attractive. For larger petals it is advisable to buy large carrots from which large pieces can be sliced either lengthwise or in rounds for the petals.

The Spider Chrysanthemum only looks complicated, and it makes an entertaining conversation piece.

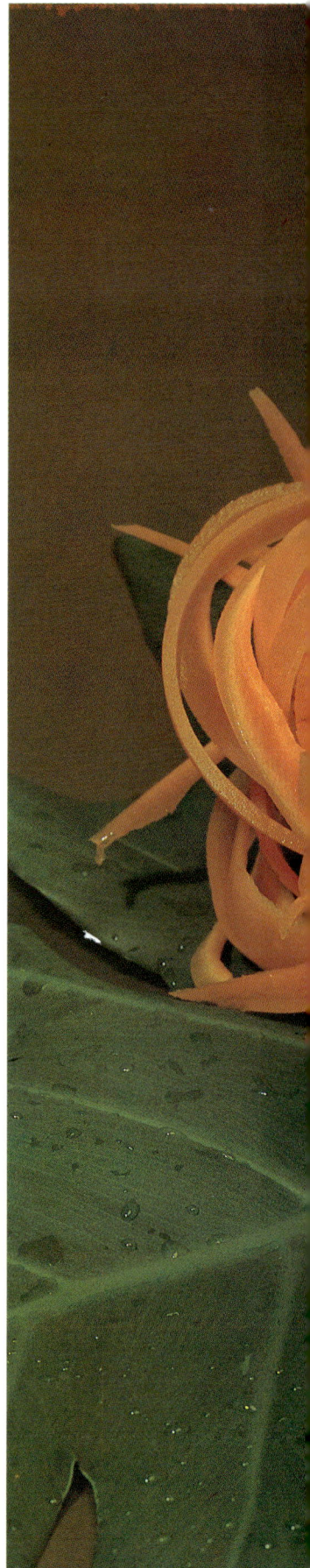

SPIDER CHRYSANTHEMUM

Select a large carrot about 20 cm long. Scrape off the skin and wash the carrot. Slice six long pieces about 2 mm thick and arrange them according to length. Take each petal and cut exactly as shown in **Figure A. Figure B** shows how the petals are to be arranged. Attach a stopper to a cocktail stick, then insert the broad end of the first petal on top of the stopper. Hold the tapering end of the petal and give it a twist before inserting that on the cocktail stick too. Start with the longest petal and proceed according to length. When all the petals have been arranged, hold them in place with a ball scooped out from a radish or carrot. Insert the other end of the cocktail stick into a cypress grass stem and use for flower arrangement. Immerse the completed flower in ice-cold water.

As this is a large, showy flower, just one flower on a broad leaf or in a pretty bowl makes a lovely centre- (and conversation) piece for the dining table.

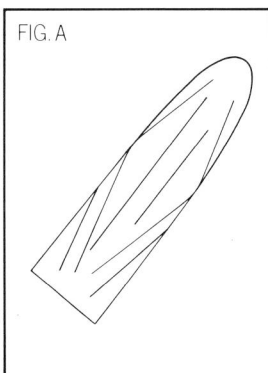

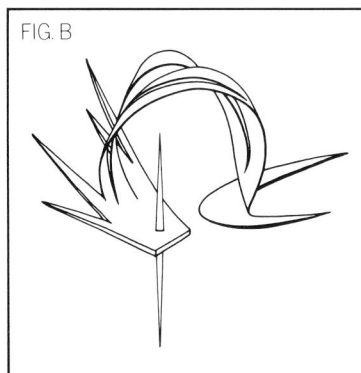

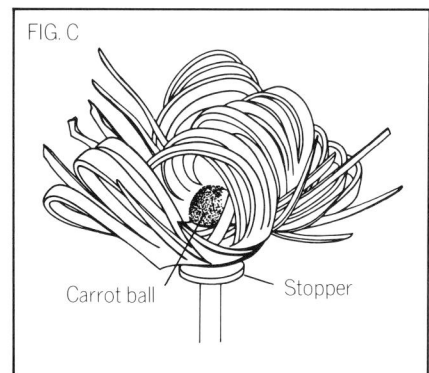

FIG. A

FIG. B

FIG. C

Carrot ball   Stopper

DAFFODIL

Illustrated on page 59

Select a large fresh carrot. Scrape away the skin and wash the carrot thoroughly. Using a knife, cut out a long piece about 7½ cm long from the broader end. From this length of carrot cut six long slices of about 3 mm thickness. Using scissors, shape the petals as in **Figure A**.

Scoop out a ball or cut a small round section from the leftover carrot and insert a cocktail stick through the centre. Then insert the base of each petal through the protruding end as shown in **Figure B**.

From the smaller end of the carrot cut out a piece about 2½ cm long. This is for the centre of the flower. Scoop out the flesh from this piece carefully and shape it so it looks like a little egg cup. Wedge the top as shown in the diagram and spike it in position over the petals.

Cut off the end of the cocktail stick protruding from the centre of the flower. Insert another cocktail stick into the stopper behind the petals so that the whole flower can be attached to a cypress stem as in the photograph on page 59.

To obtain the curved, cylindrical shape of the daffodil, put the whole flower into a cup full of cold water and refrigerate for about an hour before using for an arrangement. The cup helps the flower to retain its curved cylindrical shape.

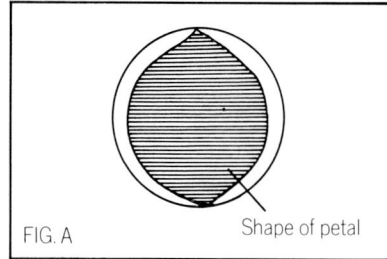

FIG. A   Shape of petal

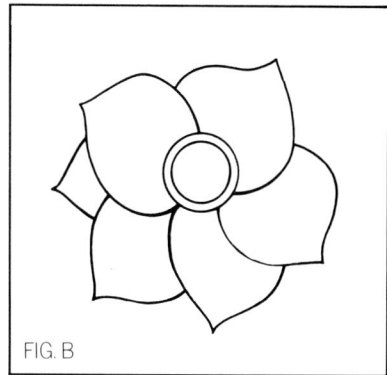

FIG. B

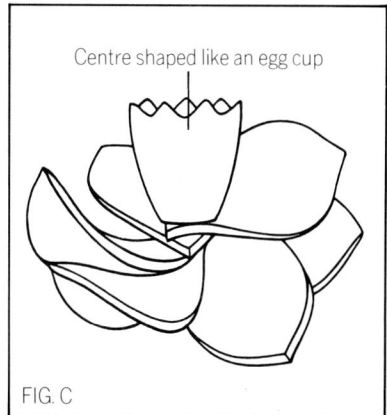

Centre shaped like an egg cup

FIG. C

FIG. D

Select a large and very fresh carrot. Scrape away the skin and wash it properly. Cut six or seven round slices from the broader end of the carrot. Each slice should be about 1 mm thick.

Insert a cocktail stick through a small stopper scooped out from leftover carrot. Spike each slice into position. The petals are spiked about 3 mm from the edge and not through the centre. See **Figure A.** Scoop out a ball from a white radish and attach it to the cocktail stick in the centre of the flower.

Carefully bend the three topmost petals one by one towards the central ball and fasten these on to the ball with toothpicks as in **Figure B.** This is to make the petals curl. Leave the flower in ice-cold water for 30 minutes or more. Before using it, remove the toothpicks.

If you make this flower with a really fresh carrot, you will see how the carrot petals curl like real flower petals. A white radish can also be used to make this flower.

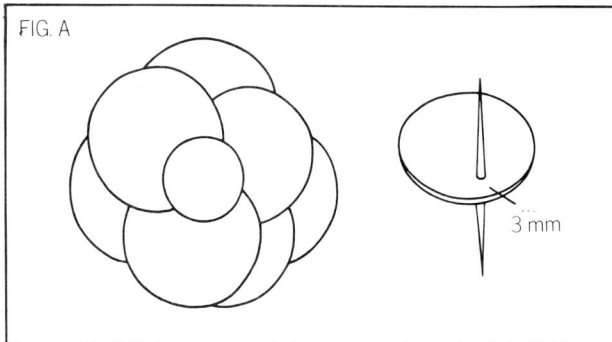

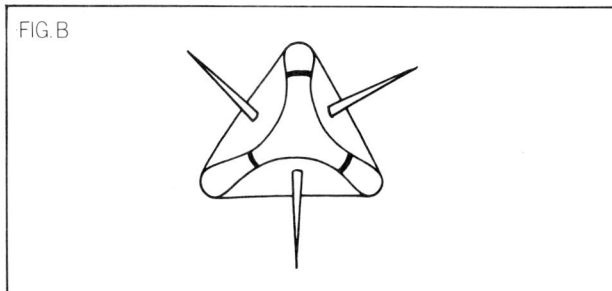

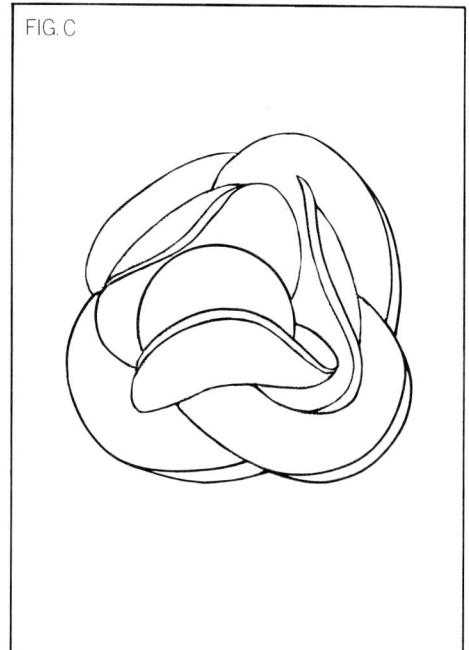

FIG. A

3 mm

FIG. B

FIG. C

Stoneware combined with the full curves of Poppies give this arrangement a charming medieval air.

GINGER FLOWER

Illustrated opposite

Scrape a small carrot to get a smooth surface. Cut off the root portion and wash thoroughly. Carve small round petals all around at the base. With the tip of a small knife, carefully remove the flesh immediately under the layer of petals. Carve the next and subsequent layers in this way till you reach the tip.

Make many ginger flowers for a large arrangement. Long leaves like New Zealand flax and cypress stems for flower stalks complement these flowers.

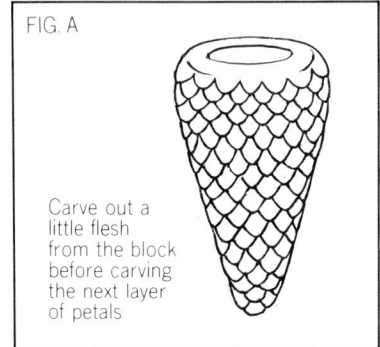

FIG. A

Carve out a little flesh from the block before carving the next layer of petals

CARROT ROSE

Illustrated opposite

The same method (as for the ginger flower) is adopted to carve the carrot rose, with a small difference. The carrot rose requires only a 4-5 cm length of carrot. Slice the base so the flower can stand. The petals should be slightly bigger than those of the ginger flower. When the centre is reached, carve a small ball. The carrot rose is an excellent food garnish.

Roses can be made from white radishes, turnips and red radishes following the same instructions. The rose on the gift-wrapped parcel in the photograph on page 21 is a turnip rose carved in this fashion and dyed yellow.

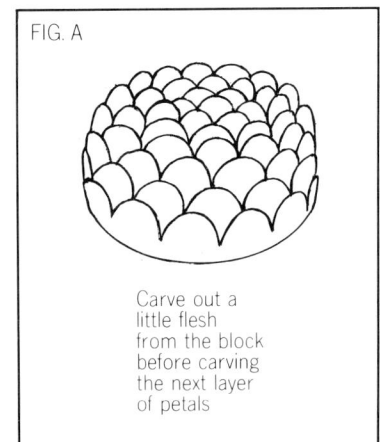

FIG. A

Carve out a little flesh from the block before carving the next layer of petals

Like an extraterrestrial seascape: driftwood and rice 'sand', outsize Ginger Flowers, a Carrot Rose and Chilli Daisies.

8 CUCUMBER

This common vegetable is mostly eaten raw, dipped in a sauce or mixed in a salad. Cucumbers are easily carved in many different patterns and are especially useful as a decoration on cooked food. Best for carving are young cucumbers which have an attractive green colour.

CUCUMBER PATTERN 1

For these decorative pieces, you will need a large green cucumber. Cut lengths of about 7½ cm. With each length, cut across diagonally so that you get two triangular pieces as in **Figures A** and **B** below. Next, make shallow V-cuts on three surfaces; the top (angled out), the back (with the skin on), and the base, as indicated in **Figure C** right. Cut thin vertical slices about 3 mm thick from this carved decorative cucumber and arrange along the edge of a dish in any desired pattern. Use any of the carved vegetable flowers with pieces of the carved cucumber to decorate cooked food.

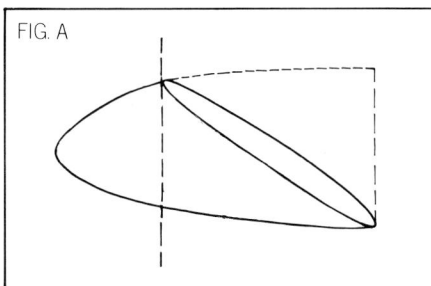

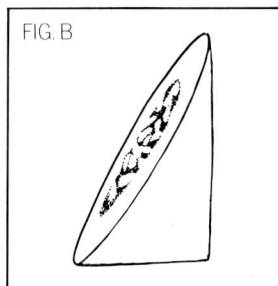

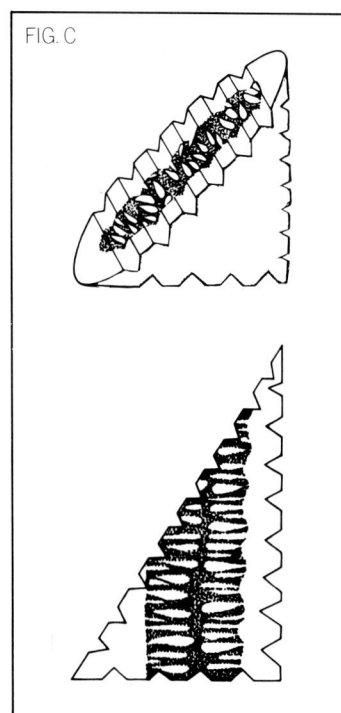

FIG. C

FIG. A

FIG. B

CUCUMBER PATTERN 2

Select a rich green fairly large cucumber. Cut across the vegetable to obtain round portions of about 3 cm length as shown in **Figure A.** Hold each portion in one hand and, using a sharp knife, carve two curved horns on both sides as in **Figure A.** Wedge the central piece as in **Figure B.** Cut the base of the flower so that it stands upright when placed on a dish. Or slice laterally to 3 mm thickness to obtain ten slices of the same pattern to decorate the edge of a dish.

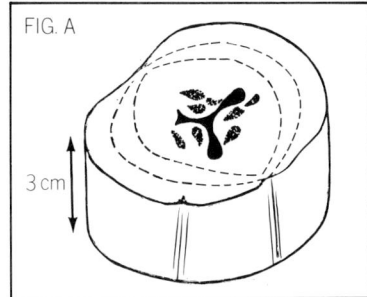
FIG. A
3 cm

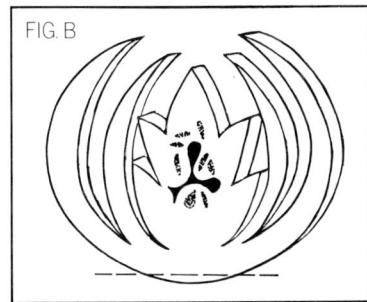
FIG. B

CUCUMBER FINGERS

You will need a large cucumber. Halve it lengthwise. Take one half and make an oblique cut right across, as shown by the broken line in **Figure A.**

Starting from this cut end, make nine thin slices, one side of which is not cut away entirely (see **Figure B**). On the ninth cut, totally slice away the finger piece from the rest of the vegetable and it will now appear as illustrated in **Figure C.** Carefully fold all the even slices between the odd ones which will form the 'fingers' (**Figure D**).

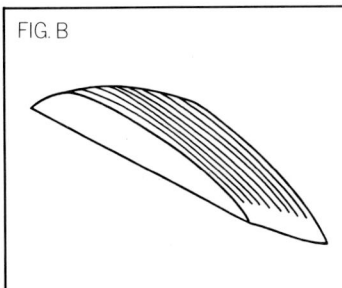
FIG. A

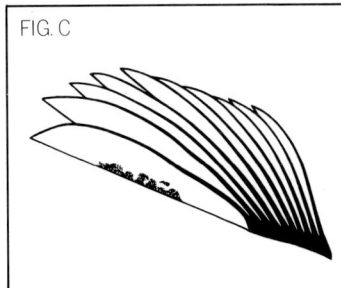
FIG. D

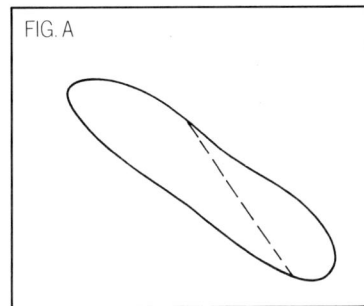
FIG. B

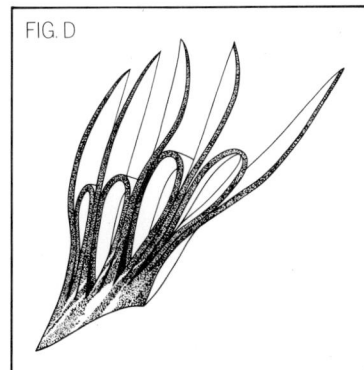
FIG. C

9 CHILLI

There are many different types of chilli but the variety most suitable for carving is the long red chilli. As they are used daily in this region for cooking curries, hot dishes and salads, they are readily available in the market. Care must be taken not to rub the eye when carving chillies as this will cause much pain and prolonged irritation.

RED BELLS

Illustrated on page 61

Select half a dozen large, red chillies and cut off the top half. Make four cuts on the sides of the chillies then trim down to the stem to form the petals. Leave the seeds in the centre. Immerse the flowers in a solution of one teaspoon sugar to one cup water. This will make the petals curl outwards quickly like a natural flower.

Use thin bamboo sticks or the sticks from a coconut-palm leaf of any desired length. Sharpen one end of each stick and attach a red bell to it. The sticks may be sheathed with stems of spring onion of the same length. Alternatively, use cypress stems.

Arrange the flowers in a vase, or use them to decorate cooked food.

FIG. A

JAPANESE WITCH–HAZEL

Illustrated on page 61

Select a red chilli about 5-7 cm in length. Using a sharp knife, slice the chilli lengthwise into two from the pointed end, to about 6 mm from the stem. Then, placing each half in turn on a cutting board, cut four thinner strips from each half. Altogether, you will have eight long strips.

The seeds are not removed as they look attractive. Immerse the flower in ice-cold water in a cup and keep the whole flower in the refrigerator. After an hour the petals will curl.

Witch-hazels are ideal to decorate fried rice, noodles, or any fish and meat dish. After use store the flower in a polythene bag and keep it in the refrigerator. The flower may be used two or three times.

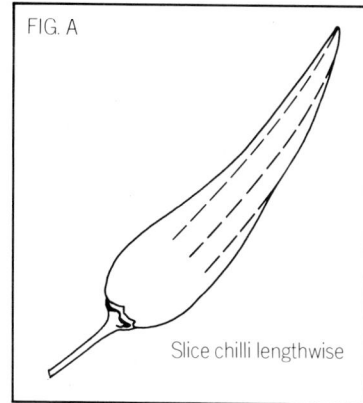

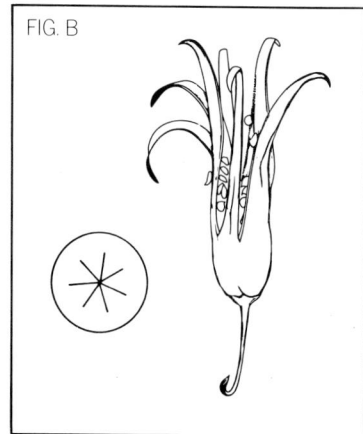

FIG. A

Slice chilli lengthwise

FIG. B

CHILLI DAISY

Illustrated on pages 61 and 69

Select a fat red chilli for this flower. Cut off the tip, leaving about 5 cm from the stalk. Proceed cutting the petals as for the witch-hazel, but trim the inside to a short 1 cm stump. Soak the flower in ice-cold water and keep refrigerated till the petals curl. Do not keep in water for more than 24 hours.

10 CAPSICUM

The green pepper or capsicum (**Figure A**) is a hollow red or green fruit and a very popular ingredient in certain Asian dishes. The fresh fruit is firm to the touch and is best for carving. Both red and green peppers may be used for carving depending on individual preference.

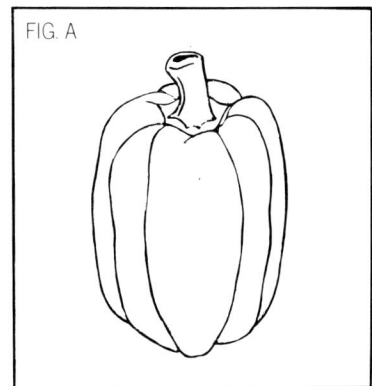

FIG. A

CAPSICUM DAISY

Illustrated on page 77

Cut off the top of a capsicum, leaving a portion about 6 cm from the bottom. Cut many vertical strips, each about ½ cm wide. Soak the flower in ice-cold water and refrigerate about an hour before using.

CAPSICUM DISH

Illustrated on page 77

Select a large green or red capsicum. Cut the 'dish' to the shape given in **Figure A** and then remove the seeds inside. The brim of the 'dish' may be shaped in a variety of patterns as desired. Check that the 'dish' stands well so that when filled it does not topple over. This makes an attractive container for serving sauces and pickles.

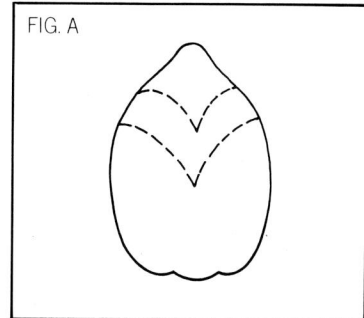

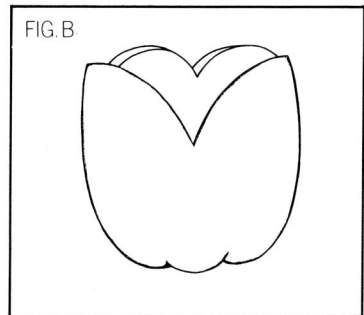

FIG. A

FIG. B

FIRE LILY

Illustrated on page 77

Select a large, red capsicum and cut off the stem portion from the top. Cut five or six petals as in the figure. These have to be shaped with a pair of scissors and pieces of the vegetable between the petals removed. In the figure, the flesh inside is removed from the stem while the seeds are left as the centre of the flower. They may be discarded if you so wish. Choose a tall capsicum to get a flower with longer petals. The Fire Lily is ideal for decorating a dish of fried noodles, fried rice or curry.

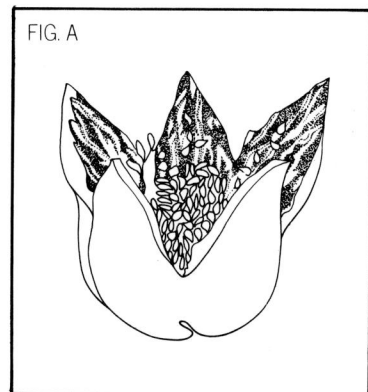

FIG. A

11 PAPAYA

Like the pineapple, the papaya or pawpaw is a very common and popular tropical fruit in Southeast Asia. The ripe fruit is often eaten raw while the half-ripe or unripe fruit is pickled or used as a vegetable in local cooking. Papaya plants bear fruit all the year round in this part of the world and the fruit is not expensive. The skin is thin and the flesh is soft especially when ripe. The fruit is therefore easy to carve in a variety of shapes and patterns.

Containers galore. Clockwise from the Radish Cup in the foreground, an Orange Lotus, Capsicum Fire-Lily, Grapefruit Dish, Papaya Container, Papaya Vase with a Capsicum Daisy, Orange Basket and Capsicum Dish.

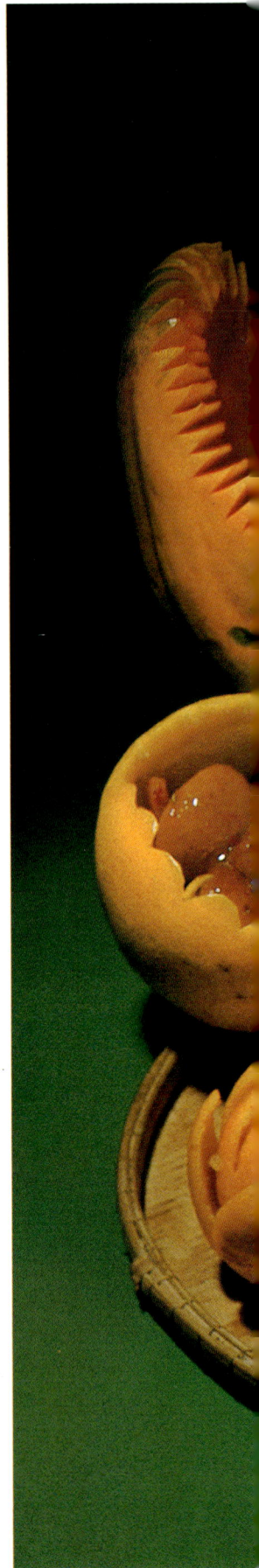

PAPAYA BASKET

Illustrated on page 77

Select a half-ripe but fairly large papaya and slice off a small section about 2 mm thick from the bottom so that it rests firmly on a serving dish. Carve two wedges from the top half of the fruit, leaving a strip about 12 mm wide as the handle of the basket. However, as the handle will be part of the fruit itself, do not select an over-ripe fruit. Remove the seeds from the inside of the fruit. This basket will be useful to serve puddings or fruit salads.

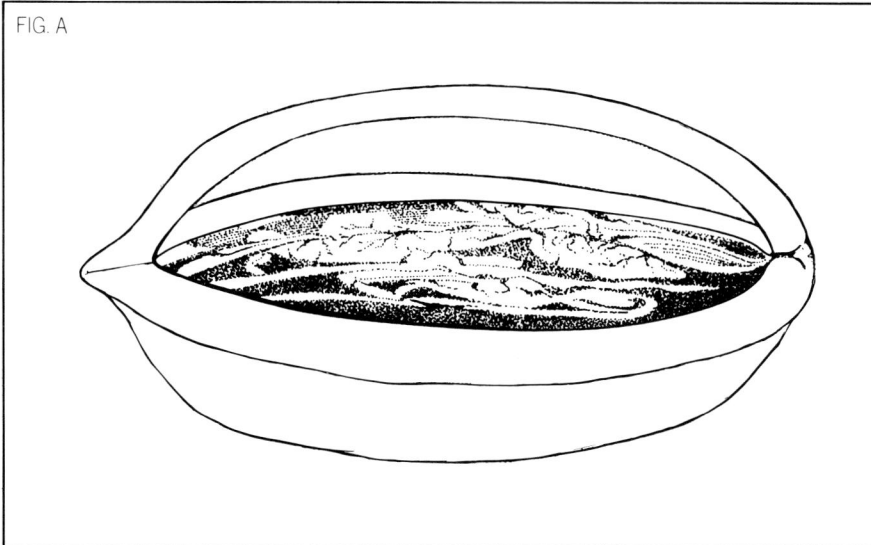

FIG. A

PAPAYA VASE

Illustrated on page 77

Select a ripe fruit which is long in shape. Slice away the top and base (**Figure A**) and use the middle section, about 25 to 30 cm in height, for the vase. Stand the vase in a dish or tray. Carve patterns or designs around the top of the vase as shown in **Figure B**. A variety of interesting designs and patterns can be carved around the side of the papaya vase as shown in the diagram. The vase is ideal for vegetable flowers attached to long stems.

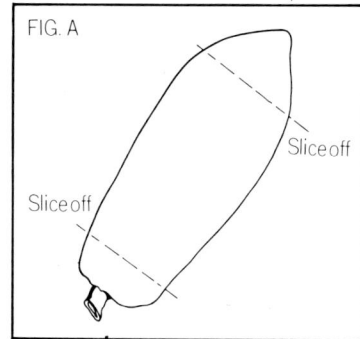

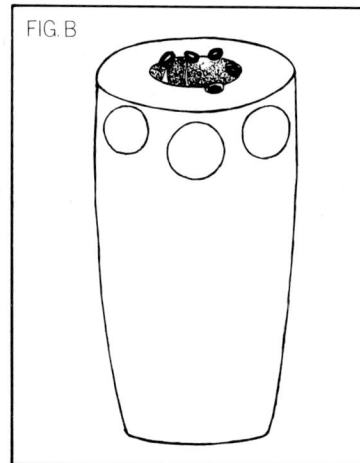

FIG. A

Slice off

Slice off

FIG. B

PAPAYA BOAT

Illustrated on page 47

As a simple exercise, try carving a papaya boat for a children's party. The result might be similar to the boat in the photograph on page 47, or a completely new creation. Cut lifeboats or flags out of carrots or radishes, and remember not to use too ripe a papaya.

12 ORANGE

The skin of the orange need not be discarded, as it usually is. Orange skins can be made into decorative flowers and containers.

ORANGE BASKET

Illustrated on page 77

You will need a few large oranges depending upon the number of guests you have. With each orange, cut off two wedges from either side as shown in **Figure A**. Remove the pulp, leaving the skin and handle in the form of a little basket. Wedge the edge of the basket in any desired pattern (see **Figure B**). This basket is used to serve puddings, fruit salads or jellies of different colours. It can be made with a large firm tomato.

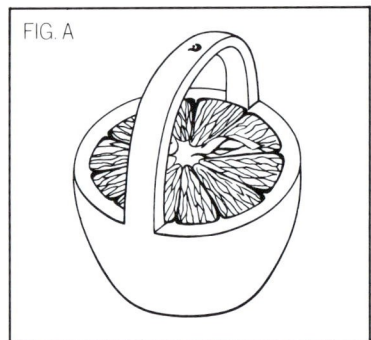

FIG. A

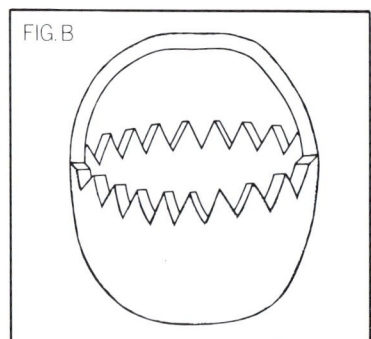

FIG. B

ORANGE LOTUS

Illustrated on page 77

For each flower, you will need the skins of three oranges. Starting from the top of one fruit, make eight equally-spaced cuts reaching nearly to the bottom. Remove the pulp to obtain one layer with eight petals. Do the same with a second fruit but make seven cuts to obtain seven petals. Trim away the pointed ends of the petals slightly so that they are shorter than the first set of petals. Next, halve the third orange laterally. Similarly remove the flesh and cut six petals on each half of the fruit. You must also trim the pointed ends of the petals.

Altogether, you will have one piece with eight petals, one with seven petals and two with six petals. Scoop out a carrot ball. With a cocktail stick, attach the ball to the centre of the smallest set of petals. Run the cocktail stick through the centres of the other sets till the largest is in. Insert a small wedge of carrot as a stopper and cut off the stick protruding from the bottom so that the lotus sits nicely on a dish. Arrange the sets so that petals of each set appear between petals of adjacent sets.

The orange lotus may be used to decorate cooked dishes or for flower arrangement.

FIG. A

13 PINEAPPLE

Pineapples are a well-known fruit and are used popularly in local cooking. The ripe fruit is very juicy and eaten raw or in a fruit cocktail as dessert. They are plentiful in the local markets all year round and are fairly cheap. The fruit, with its crown and base slips and its patterned skin, is very attractive and because it is also easy to carve, it lends itself very well to vegetable carving.

Just the thing for an Easter party—a Pineapple Container and a Pineapple Chicken with little dough chicks. Note the two Radish Roses on the handle of the basket.

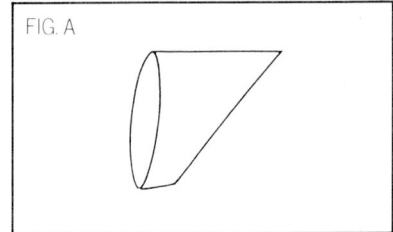
FIG. A

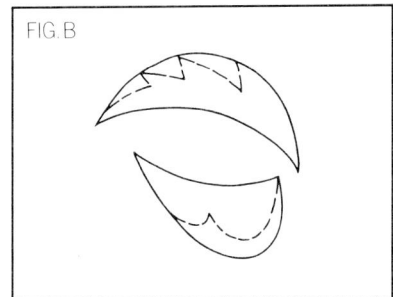
FIG. B

FIG. C

Select a pineapple which is just ripe enough for eating as a dessert and not soft. The crown of the pineapple should not be damaged or crushed as it is to resemble the tail feathers of the chicken. Slice off a small section from the 'bottom' of the pineapple so that it rests firmly on a flat surface and does not roll. Make cuts on both sides to form the wings but do not slice away completely. Cut two wedge-shaped pieces from a carrot or white radish of about 2 cm thickness as in **Figure A**, and push one under each wing of the chicken to open out the wings. Next, take a curved piece of white radish about 10 cm long to form the neck. Using cocktail sticks, attach the neck onto the pineapple as in **Figure C**. From the same vegetable, carve out the head, comb, and beak, as well as the flap under the beak, and place them in position with pins. You may dye the comb, beak and flap red with food colouring. As an alternative, cut the comb from a chilli and use the narrow end of the chilli for the chicken's beak. **Figure B** shows how the radish comb is cut. For the eyes, insert a clove on each side of the head.

The pineapple chicken is very attractive as a table decoration, especially when arranged with other vegetables, leaves or in a basket. It may also be cut up and eaten as dessert.

FIG. D

PINEAPPLE CONTAINER

Illustrated on page 83

The pineapple container, as the name suggests, is used to serve puddings, fruit cocktails and jellies. It is a decorative item.

To make the container, choose a ripe pineapple. The container is made in the same way as the pineapple chicken. However, for this item, wings are not necessary. Cut off a section from the part of the pineapple which is to be the top of the container and keep this as the cover (see **Figure A**). Cut a small piece from a carrot for a knob and with a cocktail stick or pin attach this to the cut section, to be the handle of the cover. Then scoop out the flesh from the pineapple. Make a round of plasticine and use this as the base to place the pineapple container so that it stands firmly and does not roll. (See **Figure A**.)

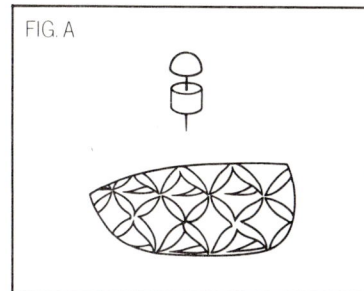

FIG. A

FIG. B

14 WATERMELON

Watermelons vary greatly in size, the average size being from about 25 to 30 cm in diameter. They are either round or oblong. The flesh is very juicy and quenches the thirst so it is often eaten raw. The skin of the melon is smooth and under it is a layer of firm white flesh, about 1 cm or more thick. Beneath this layer is the juicy red or yellow flesh. Watermelons are good for carving and ideal as containers.

White and purple Clematis flowers sitting pretty amongst a selection of melon baskets.

MELON BASKET

Illustrated on page 87

Select a ripe melon about 25 cm in diameter. Slice off a small portion of about 1 cm thickness from the bottom so that the fruit stands firmly on a dish, a large leaf or a wooden base. Begin cutting out two large wedges from either side of the top part of the fruit, leaving a band about 2 cm in width to act as the handle of the basket.

Next, use a scoop and scoop out the flesh of the fruit, including the flesh under the handle, leaving the white flesh and skin to form a basket.

Now, carve around the edge of the basket in any desired pattern or shape like the example in the diagram. Designs or patterns may also be carved on the skin of the melon.

The melon basket is commonly filled with melon balls, cold fruit cocktail, or fruit in season for dessert. Therefore, the choice of the size of the melon would also depend on the number of servings it should contain.

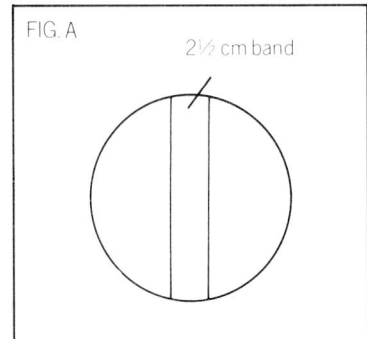

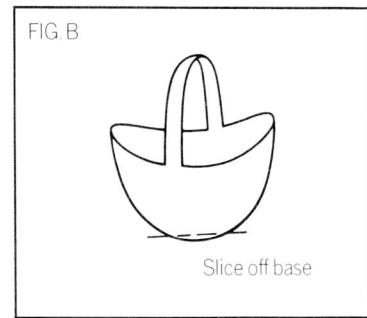

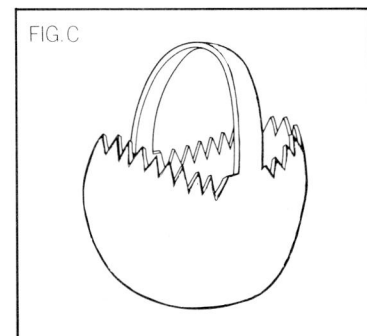

FIG. A

2½ cm band

FIG. B

Slice off base

FIG. C

15 GRAPEFRUIT

Grapefruit are large citrus fruit and yellow in colour. Popular both as a fruit and as juice for breakfast, grapefruit are available in supermarkets and some local markets. For vegetable carving buy those which are firm with bright yellow rinds and not bruised or squashed.

GRAPEFRUIT DISH

Illustrated on pages 22 and 77

Select a large grapefruit. Using a sharp knife, carve the grapefruit by shaping the rind around the top as shown in **Figure A**. A variety of patterns may be cut around the brim of the dish. Remove the flesh, taking care not to cut through the skin (see **Figure B**). The juice of the fruit may be served as a drink. The brim of the dish may be decorated tastefully with stickers.

Grapefruit dishes can be used to serve puddings, sauces or pickles.

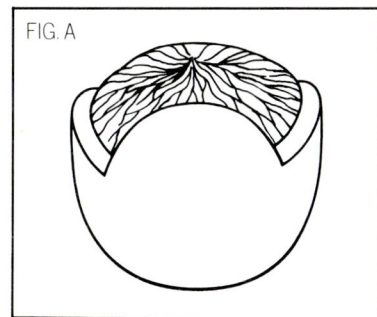

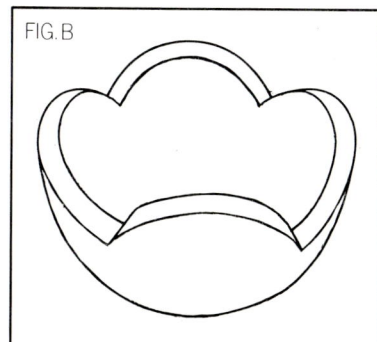

FIG. A

FIG. B

PRACTICAL HINTS

In this book, easy-to-follow steps are given on how to carve a variety of flowers from common local vegetables and copious diagrams have been included to aid the learner follow the instructions. Colourful pictures display a range of possible arrangements and uses of the flowers carved. To be successful, however, it would be wise to bear in mind some practical hints on the care and handling of equipment and the choice and buying of vegetables.

BASIC EQUIPMENT

Simple tools which can be acquired easily have been suggested as basic equipment for the vegetable carver. While these would suffice for carving the vegetables included in this book, a larger variety of equipment for special functions may be added as the carver develops greater skill in the art and wishes to carve more intricate and varied patterns of flowers.

Basic Equipment: 1 Cypress stem and leaves, 2 Cocktail sticks, 3 Slicer, 4 Brush, 5 Serrated knife, 6 Paring knife, 7 Scissors, 8 Scoop.

Cocktail Sticks

Cocktail sticks are very useful in vegetable carving especially when flowers are to be arranged in a vase and have to be attached to stalks. They are also essential for attaching petals to a stopper to keep them in place. Cocktail sticks with both ends sharpened should be obtained as in many cases one end of the stick is attached to the flower while the other end is attached to the stalk.

Containers and Pin Holders

For flower arrangement the vegetable carver will need to have a collection of pin holders and various types of containers. Containers may be low flower bowls made of glass or porcelain. However, substitutes like dishes or bowls, saucers or ashtrays may be used. The other type of container is the vase. Vases come in a large variety of designs, colours and qualities and are obtainable in many shops. In addition other containers like bamboo, driftwood, or coconut shells give originality to an. arrangement of flowers.

Pin holders are cheap and always useful in flower arrangement. Commercially produced ones with a heavy lead base and brass spikes are best but chicken wire, banana stems and even potato can be used to hold flowers in place.

Paring Knife

In vegetable carving the paring knife is used mostly for carving the petals and making clean incisions into vegetables like radishes and tomatoes. It is also useful for cutting wedges in vegetables. Select a paring knife which has a stainless steel blade about 7 cm long and is pointed at the end. The handle should be made of strong hardwood or plastic.

Potato Peeler

This is a common kitchen tool and is used to peel skins of root or fruit vegetables as suggested for some of the flowers in the book. The metal part of the peeler should preferably be of stainless steel and the cutting edge should be sharp so that the surface of the vegetable will be smooth after the skin is peeled.

Scissors or Trimmer

A pair of scissors or a trimmer is essential for vegetable carving. They are used mainly to trim petals or leaves, to cut away protruding ends of cocktail sticks and other unwanted parts of a vegetable.

The scissors or trimmer should be a medium-sized one with blades about 15 cm long. The handles may be of steel or wood. They should preferably be cutting scissors and have sharp blades. It is best to have stainless steel scissors as they are easier to clean and they do not rust. For ease of use and free movement, the scissors or trimmer should be greased with a small drop of vegetable oil periodically.

Scoop

The scoop is used to scoop out flesh from vegetables to form the central portions or stoppers of flowers. It is also used to scoop flesh from the various 'baskets' such as the papaya basket and melon basket.

Slicer

A plastic slicer with a steel blade is desirable. Since the blade is thin and extremely sharp, great care should be taken in handling it. The vegetable should be quite dry and held with a small towel over it as an extra precaution before it is put through the slicer. Choose a slicer which has an adjustable blade. The thickness of the slice may be achieved by adjusting the gap of the blade.

Utility/Serrated Knife

Like the paring knife, this knife should have a good stainless steel blade with a serrated cutting edge. It is particularly useful for slicing tomatoes or cutting large pieces and sections of carrot, cucumber, turnip and white radish. This cutting knife is much longer than the paring knife and has a blade about 18 cm long.

CARE OF EQUIPMENT

In order to avoid regular replacement as well as to derive the best results in vegetable carving, it is important to keep all equipment clean and in good condition at all times.

1 Clean knives, slicer, scoop and other metal equipment by dipping the metal parts in warm, soapy water immediately after use. Avoid immersing wooden handles or parts in the water as this would loosen or discolour them and cause screws or rivets holding them to rust. Rinse in cold water.

2 Dry and air them and arrange in a tray before storing so that they are ready for use again.

3 Use the right equipment for each particular job.

4 Keep blades of knives, scissors and slicer sharp for best results.

CHOICE AND STORAGE

The proper choice and storage of root and fruit vegetables is an important consideration in vegetable carving.

1 Generally, small or medium-sized vegetables are best as they are young and retain their shape after they are carved.

2 The vegetables chosen should be well shaped or as suggested in the instructions. They should be firm to the touch and clean with no breaks or bruises in the skin which will spoil the appearance of the carvings. Vegetables bought for carving should be a good colour and should look fresh and crisp, not withered or wrinkled with age. Unless specially required at a particular time, they are best bought at the peak of their season since they would be cheapest when in season.

3 It is best to use fresh vegetables for carving but should it be necessary to store vegetables for some time they should be cleaned thoroughly with water, wiped dry and placed in polythene bags. These should then be stored in the vegetable compartment of the refrigerator. If refrigerated too long before use, the vegetables may become brittle and break easily when they are carved.

PRESERVATION OF CARVED FLOWERS

Flowers carved from vegetables may last for varying periods of time and some of them may be used to decorate the home as well as cooked dishes for a number of times. Care in the preservation of these flowers would mean a more lasting use of them before they are used for cooking.

1 Most of the flowers suggested in this book may be made a day before they are displayed but a number of flowers, especially those carved from tomatoes, should preferably be made an hour or two before they are used.

2 Many of the flowers should be soaked in ice-cold water for half an hour either to make the petals spread or to curl them.

3 Store unused flowers in a polythene bag and leave them in the refrigerator.

4 Before displaying carved flowers, soak them in ice-cold water for 10-15 minutes to freshen them.

5 After use, the flowers may be put back in the polythene bags and again kept in the refrigerator till they are needed. Most flowers like the ones carved from red radish and carrot can be kept for more than a week in this way. Because of immersion in ice-cold water flowers which have been dyed should be re-dyed before they are displayed each time.